changing the way the world learns℠

To get extra value from this book for no additional cost, go to:

http://www.thomson.com/duxbury.html

thomson.com is the World Wide Web site for Duxbury/ITP and is your direct source to dozens of on-line resources. *thomson.com* helps you find out about supplements, experiment with demonstration software, search for a job, and send e-mail to many of our authors. You can even preview new publications and exciting new technologies.

thomson.com: It's where you'll find us in the future.

StatConcepts:
A Visual Tour of
Statistical Ideas

H. JOSEPH NEWTON

Texas A&M University

JANE L. HARVILL

Bowling Green State University

Duxbury Press
An Imprint of Brooks/Cole Publishing Company
I(T)P® An International Thomson Publishing Company

Pacific Grove, CA • Albany, NY • Bonn • Boston • Cincinnati • Detroit • Johannesburg • London
Madrid • Melbourne • Mexico City • New York • Paris • Singapore • Tokyo • Toronto • Washington

Editor: *Curt Hinrichs*
Assistant Editor: *Cynthia Mazow*
Editorial Assistant: *Rita Jaramillo*
Project Editor: *Sandra Craig*
Print Buyer: *Stacey Weinberger*
Permissions Editor: *Peggy Meehan*
Copy Editor: *Thomas L. Briggs*
Cover: *Stuart D. Paterson, Image House Inc.*
Signing Representative: *Ragu Raghavan*
Compositor: *SuperScript*
Printer: *Malloy Lithographing*

COPYRIGHT © 1997 by Brooks/Cole Publishing Company
A Division of International Thomson Publishing Inc.
I(T)P® The ITP logo is a registered trademark under license.
Duxbury Press and the leaf logo are trademarks used under license.

Printed in the United States of America
1 2 3 4 5 6 7 8 9 10

For more information, contact Duxbury Press at Brooks/Cole Publishing Company, 511 Forest Lodge Road, Pacific Grove, CA 93950, or electronically at http://www.thomson.com/duxbury.html

International Thomson Publishing Europe
Berkshire House 168–173
High Holborn
London, WC1V7AA, England

Thomas Nelson Australia
102 Dodds Street
South Melbourne 3205
Victoria, Australia

Nelson Canada
1120 Birchmount Road
Scarborough, Ontario
Canada M1K 5G4

International Thomson Publishing GmbH
Königswinterer Strasse 418
53227 Bonn, Germany

International Thomson Editores
Campos Eliseos 385, Piso 7
Col. Polanco
11560 México D.F. México

International Thomson Publishing Asia
221 Henderson Road
#05-10 Henderson Building
Singapore 0315

International Thomson Publishing Japan
Hirakawacho Kyowa Building, 3F
2-2-1 Hirakawacho
Chiyoda-hu, Tokyo 102, Japan

International Thomson Publishing Southern Africa
Building 18, Constantia Park
240 Old Pretoria Road
Halfway House, 1685 South Africa

All rights reserved. No part of this work covered by the copyright hereon may be reproduced or used in any form or by any means—graphic, electronic, or mechanical, including photocopying, recording, taping, or information storage and retrieval systems—without the written permission of the publisher.

Library of Congress Cataloging-in-Publication Data

Newton, H. Joseph, 1949–
    StatConcepts : a visual tour of statistical ideas / H. Joseph Newton, Jane L. Harvill.
        p.   cm.
    Includes index.
    ISBN 0-534-26552-9
    1. Mathematical statistics—Computer-assisted instruction
I. Harvill, Jane L. II. Title.
QA276.18.N48 1997
519.5'078'553682—dc21   97-1127

To Karah, Tim, Jake, Joseph, Katherine, and Cary

BRIEF CONTENTS

1 Introduction to StatConcepts   1

2 Random Sampling   21

3 Relative Frequency and Probability   27

4 How Are Populations Distributed?   33

5 Sampling From 0–1 Populations   57

6 Bivariate Descriptive Statistics   83

7 Central Limit Theorem   101

8 Z, t, χ^2, and F 113

9 Sampling Distributions   144

10 Minimum Variance Estimation   170

11 Interpreting Confidence Intervals   182

12 Calculating Confidence Intervals   198

13 Tests of Significance   206

14 Level of Significance of a Test   215

15 Calculating Tests of Hypotheses   244

16 Power of a Test   255

17 Calculating One-way ANOVA   289

18 Between and Within Variation   293

19 Chi-square Goodness of Fit   300

Preface xvii

Introduction and Overview of the Labs xxi

1 Introduction to StatConcepts 1

1.1 Introduction 1

1.2 Installing, Starting, and Stopping StatConcepts 1
 1.2.1 To Install StatConcepts 1
 1.2.2 To Start StatConcepts 2
 1.2.3 To Exit From StatConcepts 2

1.3 Overview of StataQuest and StatConcepts 2
 1.3.1 Basic Use of StataQuest 2
 1.3.2 Some Basic Rules in StataQuest 4
 1.3.3 Getting On-line Help 5
 1.3.4 The Data Editor 5
 1.3.5 The Nine StataQuest Windows 6
 1.3.6 The Tool Bar 7
 1.3.7 The Menu Items 8

1.4 Using the Computer As a Statistical Concepts Laboratory 17
 1.4.1 Samples From a Uniform(0,1) Population 18
 1.4.2 Random Integer From 1 to N 19
 1.4.3 Sampling With Replacement From a Finite Population 19
 1.4.4 Sampling Without Replacement From a Finite Population 19
 1.4.5 Sampling From a Continuous Population 19

1.5 The *Introduction to Concept Labs* Lab 20

2 Random Sampling 21

- **2.1** Introduction 21
  - **2.1.1** Some Basic Ideas 21
- **2.2** Objectives 22
- **2.3** Description 22
- **2.4** Guided Tour of the Lab 24
  - **2.4.1** What Does Randomness Look Like? 24
  - **2.4.2** The Behavior of Sample Means 24
- **2.5** Summary 25
- **2.6** Lab Exercises 26

3 Relative Frequency and Probability 27

- **3.1** Introduction 27
  - **3.1.1** Some Basic Ideas 27
- **3.2** Objectives 28
- **3.3** Description 28
- **3.4** Guided Tour of the Lab 30
- **3.5** Summary 32
- **3.6** Lab Exercises 32

4 How Are Populations Distributed? 33

- **4.1** Introduction 33
  - **4.1.1** Some Basic Ideas 33
- **4.2** Objectives 34
- **4.3** Description 34
- **4.4** Guided Tour of the Lab 35
  - **4.4.1** The Normal Family 35
  - **4.4.2** The Student t Family 37
  - **4.4.3** The Chi-square Family 38
  - **4.4.4** The F Family 40
  - **4.4.5** The Beta Family 41
  - **4.4.6** The Cauchy Family 42
  - **4.4.7** The Exponential Family 44
  - **4.4.8** The Gamma Family 45
  - **4.4.9** The Laplace Family 46

	4.4.10	The Logistic Family 46
	4.4.11	The Lognormal Family 47
	4.4.12	The Pareto Family 48
	4.4.13	The Uniform Family 50
	4.4.14	The Weibull Family 51
4.5	Summary 53	
4.6	Lab Exercises 53	
4.7	Formulas for Continuous Distributions, Means, and Variances 54	

5 Sampling From 0–1 Populations 57

5.1 Introduction 57
  5.1.1 Some Basic Ideas 57

5.2 Objectives 59

5.3 Description 60
  5.3.1 Sampling With and Without Replacement 60
  5.3.2 The Negative Binomial Distribution 61
  5.3.3 Approximating Binomial Probabilities 62

5.4 Guided Tour of the Lab 65
  5.4.1 Sampling With and Without Replacement 65
  5.4.2 The Negative Binomial Distribution 70
  5.4.3 Approximating Binomial Probabilities 72

5.5 Summary 78

5.6 Lab Exercises 79
  5.6.1 Sampling With and Without Replacement 79
  5.6.2 The Negative Binomial Distribution 80
  5.6.3 Approximating Binomial Probabilities 80

5.7 Formulas for Discrete Distributions, Means, and Variances 81

6 Bivariate Descriptive Statistics 83

6.1 Introduction 83
  6.1.1 Two Sampling Schemes for Bivariate Populations 83
  6.1.2 Scatterplots 84
  6.1.3 Correlation Coefficient 84
  6.1.4 Least Squares Regression Line 86
  6.1.5 Residuals and the Multiple Correlation Coefficient, r^2 87

6.2 Objectives 88

6.3 Description 88
- **6.3.1** Scatterplots I 88
- **6.3.2** Scatterplots II 89
- **6.3.3** Least Squares 90

6.4 Guided Tour of the Lab 92
- **6.4.1** Scatterplots I 92
- **6.4.2** Scatterplots II 93
- **6.4.3** Least Squares 94

6.5 Summary 99

6.6 Lab Exercises 99
- **6.6.1** Scatterplots I 99
- **6.6.2** Scatterplots II 99
- **6.6.3** Least Squares 100

7 Central Limit Theorem 101

7.1 Introduction 101
- **7.1.1** Some Basic Ideas 101

7.2 Objectives 102

7.3 Description 102

7.4 Guided Tour of the Lab 105
- **7.4.1** Parent-Curves 105
- **7.4.2** One-at-a-time 106
- **7.4.3** 500-Samples 109

7.5 Summary 110

7.6 Lab Exercises 111
- **7.6.1** Parent-Curves 111
- **7.6.2** One-at-a-time 111
- **7.6.3** 500-Samples 112

8 Z, t, χ^2, and F 113

8.1 Introduction 113
- **8.1.1** Some Basic Ideas 113
- **8.1.2** The Normal Family 114
- **8.1.3** The t Family 114
- **8.1.4** The χ^2 Family 114
- **8.1.5** The F Family 114
- **8.1.6** Critical Values 115

8.2 Objectives 115

- **8.3** Description 116
  - **8.3.1** Critical Values 116
  - **8.3.2** Normal Curves 118
  - **8.3.3** Chi-square Curves 119
  - **8.3.4** F Curves 121
  - **8.3.5** t Converging to Z 122
  - **8.3.6** Normal Approximation to Binomial 123

- **8.4** Guided Tour of the Lab 125
  - **8.4.1** Critical Values 125
  - **8.4.2** Normal Curves 129
  - **8.4.3** Chi-square Curves 130
  - **8.4.4** F Curves 132
  - **8.4.5** t Converging to Z 134
  - **8.4.6** Normal Approximation to Binomial 136

- **8.5** Summary 138

- **8.6** Lab Exercises 139
  - **8.6.1** Critical Values 139
  - **8.6.2** Normal Curves 140
  - **8.6.3** Chi-square Curves 141
  - **8.6.4** F Curves 142
  - **8.6.5** t Converging to Z 143
  - **8.6.6** Normal Approximation to Binomial 143

9 Sampling Distributions 144

- **9.1** Introduction 144
  - **9.1.1** Some Basic Ideas 145

- **9.2** Objectives 147

- **9.3** Description 147

- **9.4** Guided Tour of the Lab 149
  - **9.4.1** The Sampling Distribution of the Z Statistic 149
  - **9.4.2** The Sampling Distribution of the One-sample t Statistic 151
  - **9.4.3** The Sampling Distribution of the Two-sample t Statistic 157
  - **9.4.4** The Sampling Distribution of the χ^2 Statistic 159
  - **9.4.5** The Sampling Distribution of the F Statistic 162

- **9.5** Summary 165

- **9.6** Lab Exercises 166
  - **9.6.1** The Sampling Distribution of the Z Statistic 166
  - **9.6.2** The Sampling Distribution of the One-sample t Statistic 166
  - **9.6.3** The Sampling Distribution of the Two-sample t Statistic 167
  - **9.6.4** The Sampling Distribution of the χ^2 Statistic 168
  - **9.6.5** The Sampling Distribution of the F Statistic 168

10 Minimum Variance Estimation   170

- **10.1** Introduction   170
  - **10.1.1** Some Basic Ideas   170
- **10.2** Objectives   172
- **10.3** Description   172
- **10.4** Guided Tour of the Lab   174
  - **10.4.1** The Normal(0,1) Population Distribution   175
  - **10.4.2** The Uniform(−0.5,0.5) Population Distribution   177
  - **10.4.3** The Laplace(0,1) Population Distribution   177
  - **10.4.4** The t_3 Population Distribution   178
- **10.5** Summary   179
- **10.6** Lab Exercises   180
  - **10.6.1** The Normal(0,1) Population Distribution   180
  - **10.6.2** The Uniform(−0.5,0.5) Population Distribution   180
  - **10.6.3** The Laplace(0,1) Population Distribution   181
  - **10.6.4** The t_3 Population Distribution   181

11 Interpreting Confidence Intervals   182

- **11.1** Introduction   182
  - **11.1.1** Some Basic Ideas   183
- **11.2** Objectives   185
- **11.3** Description   185
- **11.4** Guided Tour of the Lab   188
  - **11.4.1** The Effect of Sample Size   188
  - **11.4.2** The Effect of Confidence Level   189
  - **11.4.3** Violation of Assumptions   190
  - **11.4.4** Intervals for Other Parameters   190
- **11.5** Summary   195
- **11.6** Lab Exercises   196

12 Calculating Confidence Intervals   198

- **12.1** Introduction   198
- **12.2** Objectives   198
- **12.3** Description   198

12.4	Guided Tour of the Lab 200	
	12.4.1 Example 1 200	
	12.4.2 Example 2 200	
	12.4.3 Example 3 201	
	12.4.4 Example 4 202	
	12.4.5 Example 5 202	
	12.4.6 Example 6 203	
	12.4.7 Example 7 203	
12.5	Summary 204	
12.6	Confidence Interval Formulas 204	

13 Tests of Significance 206

13.1 Introduction 206
    13.1.1 Types of Tests 207

13.2 Objectives 209

13.3 Description 209

13.4 Guided Tour of the Lab 211
    13.4.1 Finding p Values for Two-sided Tests 211
    13.4.2 p Values As Statistics Get More Extreme 212

13.5 Summary 213

13.6 Lab Exercises 213

14 Level of Significance of a Test 215

14.1 Introduction 215
    14.1.1 Some Basic Ideas 215

14.2 Objectives 216

14.3 Description 217

14.4 Guided Tour of the Lab 219
    14.4.1 The Z Test Statistic 219
    14.4.2 The One-sample t Test Statistic 225
    14.4.3 The Two-sample t Test Statistic 228
    14.4.4 The χ^2 Test Statistic 231
    14.4.5 The F Test Statistic 234

14.5 Summary 236

14.6 Lab Exercises 237
    14.6.1 The Z Test Statistic 237
    14.6.2 The One-sample t Test Statistic 239

14.6.3 The Two-sample t Test Statistic 240
14.6.4 The χ^2 Test Statistic 241
14.6.5 The F Statistic 242

15 Calculating Tests of Hypotheses 244

15.1 Introduction 244

15.2 Objectives 244

15.3 Description 244

15.4 Guided Tour of the Lab 246
15.4.1 Example 1 246
15.4.2 Example 2 247
15.4.3 Example 3 248

15.5 Summary 250

15.6 Confidence Interval and Tests of Hypotheses Formulas 251

16 Power of a Test 255

16.1 Introduction 255
16.1.1 Some Basic Ideas 255

16.2 Objectives 258

16.3 Description 258

16.4 Guided Tour of the Lab 260
16.4.1 The Z Test Statistic 260
16.4.2 Choosing Sample Sizes 265
16.4.3 The One-sample t Test Statistic 266
16.4.4 The Two-sample t Test Statistic 272
16.4.5 The χ^2 Test Statistic 275
16.4.6 The F Test Statistic 279

16.5 Summary 283

16.6 Lab Exercises 284
16.6.1 The Z Test Statistic 284
16.6.2 The One-sample t Test Statistic 286
16.6.3 The Two-sample t Test Statistic 286
16.6.4 The χ^2 Test Statistic 287
16.6.5 The F Test Statistic 288

17 Calculating One-way ANOVA 289

- **17.1** Introduction 289
- **17.2** Objectives 290
- **17.3** Description 290
- **17.4** Guided Tour of the Lab 291
  - **17.4.1** The Data 291
  - **17.4.2** Analysis of the Example 291
- **17.5** Summary 292

18 Between and Within Variation 293

- **18.1** Introduction 293
- **18.2** Objectives 294
- **18.3** Description 294
- **18.4** Guided Tour of the Lab 296
  - **18.4.1** Level of Significance and Power of the F Test 296
  - **18.4.2** The Effect of Variation Within Samples 296
  - **18.4.3** The Effect of Sample Size 297
- **18.5** Summary 298
- **18.6** Lab Exercises 298

19 Chi-square Goodness of Fit 300

- **19.1** Introduction 300
- **19.2** Objectives 301
- **19.3** Description 301
- **19.4** Guided Tour of the Lab 302
- **19.5** Summary 303
- **19.6** Lab Exercises 304

Index 305

PREFACE

Introduction

Most introductory statistics courses have three parts: (1) *descriptive statistics*, which uses numbers and graphs to summarize the information about a data set, (2) *inferential statistics*, which draws conclusions about numerical characteristics of entire populations of objects from those of samples from the populations, and (3) *statistical concepts*, which are the basic logical and mathematical ideas underpinning descriptive and inferential statistics.

A wide variety of computer programs make it easy for students to accomplish what is required for the first two of these parts, but very little software has been developed for illustrating statistical concepts. That's why we wrote StatConcepts—as a set of "laboratories" for illustrating these concepts.

StatConcepts is actually a collection of programs written in the language of StataQuest, a student version of a program called Stata that is designed to do descriptive and inferential statistics.

StatConcepts is not intended as a text, but as a supplement to introductory statistics texts. Its main focus is on correct intrepretation and understanding of statistical concepts, terminology, and results, not on computation for a given problem. However, StatConcepts does contain some labs that allow students to compute results.

In many ways, the computer is the laboratory for the science of statistics. Most statistical investigations have their roots in a statement along these lines: "If we did this procedure over and over again, then this is what we would see." The only way realistically to do things over and over again is on a computer. In these labs, we have tried to use graphics to show what, in fact, we would see if we did various things over and over again.

We assume that instructors will not incorporate all of the labs in the StatConcepts collection (there are 28 of them!) into their courses, but rather pick and choose those they feel would be most useful in the course (and have time to cover in their already cramped schedule).

We hope that instructors can show the labs to students using some kind of projection, but each chapter of this book contains a "guided tour" through each lab that a student could read while at a computer. These guided tours cannot totally replace an instructor, but they can certainly help instructors use the labs as a supplement to their courses.

Although the labs and this book are intended primarily for introductory courses, we have found them very valuable in courses at all levels. We have kept the material as nontechnical as possible, but more advanced students will be able to relate to the graphs and descriptions at a more mathematical level.

Computer Requirements for Using StatConcepts

From a software point of view, StatConcepts is totally self-contained and requires only a computer running Microsoft Windows.

The Structure of the Chapters

Except for the introductory lab and the *Calculating Confidence Intervals*, *Calculating Tests of Hypotheses*, and *Calculating One-way ANOVA* Labs (which are more calculation than concepts oriented), each chapter in this book is structured in the same way:

1. *Introduction:* The first section provides background information needed for the lab or labs in the chapter.
2. *Objectives:* This section briefly summarizes the concepts the lab or labs illustrate.
3. *Description:* This section briefly describes how the lab or labs work and what the items in the dialog box are.
4. *Guided Tour of the Lab:* This section is the heart of each chapter. Although the labs are best used while in front of the computer, we have included enough graphs from the labs to communicate the basic ideas by simply reading the tour. Some of the tours have many stops. Again, we have tried to design them so that readers can visit as many of the stops on a tour as they have the time or interest for.
5. *Summary:* This section summarizes what the guided tour has illustrated.
6. *Lab Exercises:* This section contains a set of exercises that can be used to further illustrate the key ideas.

Acknowledgments

We have received an amazing amount of help from a wide variety of people in creating and finishing this project. First, thanks to the people at Stata Corporation for having written such a nice piece of software, particularly the graphics and dialog boxes that make StatConcepts possible. James Hardin was particularly helpful in answering our many questions.

We also appreciate the comments of the manuscript reviewers: Robert Hale, Pennsylvania State University; Robert Heckard, Pennsylvania State University;

David C. Howell, The University of Vermont; Dennis Jowaisas, Oklahoma City University; Bill Notz, Ohio State University; and Bill Seaver, University of Tennessee, Knoxville.

We are indebted to Stan Loll, Alexander Kugushev, and Curt Hinrichs of Duxbury Press for their encouragement at the beginning of the project and to Sandra Craig, William Baxter, and Thomas Briggs for their yeoman work in the production of the book.

At least 25 different instructors here at Texas A&M have used preliminary versions of these labs in teaching our large elementary statistics courses, and they have given us many wonderful suggestions. We would like to single out Naisyin Wang, Donald Lancon, Andy Liaw, and, in particular, Julie Hagen Carroll for their special help.

Finally, we thank our families, and especially Linda and Marty, for their patience and support.

H. Joseph Newton
Department of Statistics
Texas A&M University
College Station, TX 77843-3143
jnewton@stat.tamu.edu

Jane L. Harvill
Applied Statistics and Operations Research
Bowling Green State University
Bowling Green, OH 43403-0267
jharvil@cba.bgsu.edu

INTRODUCTION AND OVERVIEW OF THE LABS

As noted previously, there are 28 labs in all, although there are fewer items on the Labs menu because some items have submenus containing more than one lab. There is a chapter in this book for each item on the Labs menu, as follows:

1 *Introduction to Concept Labs:* This lab is actually just a greeting and an invitation to look at a help file giving an overview of the entire collection of labs. It also allows users to specify their own random number generator seed.

2 *Random Sampling:* This lab repeatedly shows random sampling without replacement from a population of 100 boxes. It also previews the ideas of sampling distributions and the Central Limit Theorem.

3 *Relative Frequency and Probability:* This lab again illustrates random sampling without replacement using the example of a lottery game. It also illustrates the relative frequency interpretation of probability by repeatedly drawing six winning numbers from the numbers 1–50 and keeping track of the number of draws containing at least two consecutive numbers. Deriving the formula for the probability of this event is beyond the scope of most courses.

4 *How Are Populations Distributed?* This lab shows students that distributions come in all shapes and sizes and in parametric families. The lab graphs densities from 14 different families. It also generates random samples from one member of each family and superimposes the density on the histogram of the sample, thus illustrating variability from one sample to another.

5 *Sampling From 0–1 Populations:* This item actually leads to four different labs:

   (a) *Sampling With and Without Replacement:* The binomial and hypergeometric distributions are illustrated by having the user specify the number of elements in a 0–1 population, the proportion of 1's, and the size of a sample, and then superimposing the probability plot of the number of 1's in the sample under conditions of sampling with and without replacement.

   (b) *The Negative Binomial Distribution:* This lab graphs the negative binomial distribution for user-specified values of the parameters.

   (c) *Poisson Approximation to Binomial:* This lab superimposes the binomial distribution and its Poisson approximation for user-specified values of the parameters. It makes it easy to see when the Poisson approximation works well and when it doesn't.

  (d) *Normal Approximation to Binomial:* This lab superimposes the binomial distribution and its normal approximation for user-specified values of the parameters. It makes it easy to see when the normal approximation works well and when it doesn't.

6 *Bivariate Descriptive Statistics:* This item leads to three different labs:

  (a) *Scatterplots I:* This lab shows scatterplots of random samples from a bivariate normal population for 20 different values of the correlation coefficient ranging from −0.9 to 0.9.

  (b) *Scatterplots II:* This lab allows the user to generate scatterplots for any sample size and any population correlation coefficient.

  (c) *Least Squares:* This lab allows the user to generate a wide variety of different scatterplots and then see the true line, the least squares line, and the vertical errors that go into the residual sum of squares.

7 *Central Limit Theorem:* This lab illustrates sample means for repeated sampling from a user-specified choice of four parent populations: normal, exponential, uniform, and 0–1. This is actually two labs in one:

  (a) *One-at-a-time:* One sample at a time, boxes corresponding to sample means are placed above an axis until the tallest column of boxes fills the graph.

  (b) *500 Samples:* The histogram of the sample means for 500 samples is drawn with the approximating normal curve superimposed.

8 Z, t, *Chi-square, and* F: This item leads to six labs:

  (a) *Critical Values:* This lab graphs rejection regions for one- and two-tailed tests for any of Z, t, χ^2, or F for user-specified α and, if necessary, degrees of freedom.

  (b) *Normal Curves:* This lab starts by drawing the standard normal curve. The user can then repeatedly change the mean and/or variance, and each time the lab draws the new normal curve on the same axes.

  (c) *Chi-square Curves:* This lab starts by drawing the χ^2 curve with 10 degrees of freedom. The user can then repeatedly change the degrees of freedom, and each time the lab draws the new χ^2 curve on the same axes.

  (d) *F Curves:* This lab starts by drawing the F curve with 10 and 10 degrees of freedom. The user can then repeatedly change the degrees of freedom, and each time the lab draws the new F curve on the same axes.

  (e) *t Converging to Z:* This lab allows the user to superimpose any part of the Z curve and the same part of the t curve for increasing degrees of freedom.

  (f) *Normal Approximation to Binomial:* This is the same lab as in the *Sampling From 0–1 Populations* item.

9 *Sampling Distributions:* This lab allows the user to generate 500 samples (or pairs of samples) of user-specified size from one of three parent populations: normal, uniform, and exponential. The user can then calculate the Z, one- or two-sample t, χ^2, or F statistics and superimpose the histogram of the 500 statistics and the

theoretical normal theory curve. The lab also displays the percentiles of the 500 statistics and the theoretical curve to see the agreement (disagreement) of the two if assumptions are (are not) met.

10 *Minimum Variance Estimation:* This lab allows the user to generate 500 samples of user-specified size from one of four parent populations (Normal(0,1), Uniform(−0.5, 0.5), t with 3 degrees of freedom, and Laplace), each symmetrical about zero, and then draw the histograms of the 500 sample means and 500 sample medians. The sample mean and standard deviations of the 500 means and 500 medians are also displayed. The lab shows that the sample mean is not always the best estimator.

11 *Interpreting Confidence Intervals:* This lab allows the user to generate 50, 100, or 150 samples (or pairs of samples) of user-specified size from one of four parent populations (normal, uniform, exponential, and 0–1) and draw horizontal lines for the confidence intervals (for user-specified α) for user-specified parameters (μ, σ^2, $\mu_1 - \mu_2$, σ_1^2/σ_2^2, π, or $\pi_1 - \pi_2$), as well as a vertical line representing the true value of the parameter. This allows the user to see the effect of changing α and sample size on the width of intervals, as well as the effect of violating assumptions on the confidence interval coverage probability.

12 *Calculating Confidence Intervals:* This lab allows the user to calculate confidence intervals for the 11 different one- and two-sample inference situations for means, variances, and proportion problems usually covered in an introductory course.

13 *Tests of Significance:* This lab draws a graph of a user-specified Z, t, χ^2, or F curve and shades in the area corresponding to the p value for either a user-specified or lab-generated value of a test statistic.

14 *Level of Significance of a Test:* This lab allows the user to generate 500 samples (or pairs of samples) of user-specified size from one of three parent populations (normal, uniform, and exponential); calculate the Z, one- or two-sample t, χ^2, or F statistics; and then superimpose the histogram of the 500 statistics and the theoretical normal theory curve. It shades in the area under the curve for the rejection region (for the user-specified α and one- or two-tailed test) and displays the proportion of times the null hypothesis is rejected, thus showing the agreement (disagreement) with the value of α if assumptions are (are not) met.

15 *Calculating Tests of Hypotheses:* This lab allows the user to calculate test statistics and p values for the same 11 situations as in the *Calculating Confidence Intervals* Lab. It draws a graph with the test statistic marked and the tail areas corresponding to the p value shaded in.

16 *Power of a Test:* This lab is the same as the *Level of Significance of a Test* Lab except that now the user can specify the degree to which the null hypothesis is actually false (including actually being true). This allows the user to see the effect of sample size and degree of falseness on the power of the test.

17 *Calculating One-way ANOVA:* This lab allows the user to enter sample sizes, means, and variances (or standard deviations) for specified number of samples and then displays the ANOVA table.

18. *Between and Within Variation:* This lab starts by generating and graphing two sets of four samples, each of user-specified size. In the first set of four samples, the populations have differing means; in the second set, the population means are all the same. All the populations have the same variance. Then the lab allows the user to repeatedly change the population variance, each time redrawing the plot and displaying the p value of the one-way ANOVA test of equality of means. This allows the user to see how the test is comparing the between-sample variability to the within-sample variability.

19. *Chi-square Goodness of Fit:* This lab illustrates the χ^2 goodness-of-fit test by generating a user-specified number of points in a square and then placing a grid of (user-specified number of) boxes on the points, counting how many are in each box, and then displaying the p value of the resulting χ^2 test. This allows users to see how nonuniform the placement of points can look even though they are, in fact, being placed according to a uniform distribution.

Introduction to StatConcepts

This chapter sets the stage for the detailed descriptions of the 29 labs contained in StatConcepts.

1.1 Introduction

In this chapter, we give an overview of how StatConcepts operates. Specifically, we tell you how to install, start, and stop it, as well as briefly describe all the parts of the program. We also review how experiments are actually performed on a computer, including how random number generators work and how sampling from various kinds of populations is done. Finally, we discuss the *Introduction to Concept Labs* Lab.

1.2 Installing, Starting, and Stopping StatConcepts

1.2.1 To Install StatConcepts

Make sure that Windows 95 or Windows, version 3.1 or later, is running on your computer and that you have inserted the StatConcepts diskette into the disk drive. If you are running Windows 95, you have a choice of two methods of installing StatConcepts. The first method is preferred since it allows you to uninstall the program later if you so desire. Go to the *Start* menu and choose *Settings*, then *Control Panel*, then *Add/Remove Programs*, and finally click on *Install*. Then you should follow the directions that will appear on the screen. The second installation procedure for

Windows 95 consists of going to the *Start* menu and choosing *Run* and then typing `drive:setup` where `drive` is the letter of the diskette drive you are using, and then following the instructions on the screen.

If you are running Windows 3.x, chose *Run* from the *File* menu in the Program Manager. A dialog box should appear in which you should type `drive:setup` where `drive` is the letter of the diskette drive you are using. Then click on "OK" and follow the instructions on the screen.

1.2.2 To Start StatConcepts

You start StatConcepts as you do any other program under Windows. If you are using Windows 95, StatConcepts will appear on the *Start* menu and you can start it by selecting it from that menu. You can also create a shortcut to StatConcepts and place it on your desktop. If you are running Windows 3.x, you start StatConcepts by double-clicking on its icon or by pressing ENTER while the icon is highlighted.

1.2.3 To Exit From StatConcepts

From the *File* menu, choose *Exit*. If you have analyzed data using the procedures in StataQuest and have changed the data or a label or the StataQuest procedures have sorted the data differently from its original order, StatConcepts will tell you that the data set has changed and ask you if you really want to exit. You must answer OK to exit.

1.3 Overview of StataQuest and StatConcepts

As discussed in the Preface, StatConcepts is actually StataQuest with some additional capabilities. StataQuest is an easy-to-use program for doing descriptive and inferential statistics in introductory statistics courses.

When you start StataQuest, the screen shows four windows (see Figure 1.1) labeled `Stata Results`, `Stata Command`, `Variables`, and `Review`, as well as a row of *menu items* across the top of the screen and a row of *buttons* underneath the menu items. At this point, the only difference between StatConcepts and StataQuest is the menu item in the top row labeled `Labs`.

Note that all the windows and buttons are within the overall StataQuest window, which can be resized the same way as any Microsoft Windows window.

1.3.1 Basic Use of StataQuest

In a typical StataQuest session, users read or create a data set consisting of one or more variables (such as people's height and weight) and produce some numerical (means, medians, standard deviations, and so on) or graphical (histograms, box plots, and so on) descriptive statistics and/or some statistical inferences such as confidence intervals or tests of hypotheses. They then either print the results directly or copy the numerical and graphical output into a report being produced by a word processor and print the report from the word processor.

FIGURE 1.1
The StatConcepts window when the program starts.

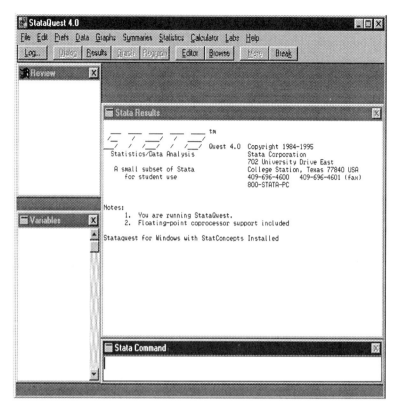

We assume you are familiar with the basic operations of Microsoft Windows—that is, how to start programs, how to resize and move windows, and how to switch back and forth from one application (such as StatConcepts) to another (such as a word processor).

All of the capabilities of StataQuest and StatConcepts can be accessed by clicking on one of the menu items at the top of the window. In many cases, doing so will bring up a submenu of further choices, and clicking on one of these items will sometimes lead to still another set of options. For example, below we show how you would form the histogram of a continuous variable:

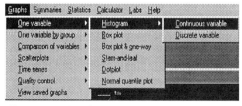

Thus, you click first on Graphs, then on One variable in the menu that appears, then on Histogram, and finally on Continuous variable. We will denote such a series of "clicks" by

Graphs→ One variable→ Histogram→ Continuous variable.

With a few exceptions, choosing one of the StataQuest items causes the program to open a *dialog box* containing some blank spaces that have to be filled in before StataQuest will actually perform the desired analysis. For example, specifying that a histogram is wanted leads to the dialog box shown in Figure 1.2.

FIGURE **1.2**
Histogram dialog box.

1.3.2 Some Basic Rules in StataQuest

Before we describe all the things that StatConcepts and StataQuest can do, here are the basic properties of the program:

1. All the data to be analyzed must be in a rectangular array called a *data set*. Each column of the array is called a *variable*, and each row is called a *case*. In StataQuest, only one data set can be *active* at a time, and each data set can have at most 25 variables. No variable can have more than 600 elements, and the individual numbers can total at most 4000.

2. Data sets have *labels*, as do individual variables. These labels are used in the output of many commands.

3. Data can be either *numerical* or *string* valued; a string variable is one that contains characters.

4. Data sets can be formed by typing them using the StataQuest *editor*, which is in the form of a standard spreadsheet program. Once the data set has been formed, it can be saved to a file for later retrieval (via File→ Open) using File→ Save. This saves the data set in StataQuest format, which cannot be read directly by other programs. To save a data set to a file in standard ASCII format that can be read by other programs, use File→ Export ASCII. You can read ASCII files using File→ Import ASCII.

1.3.3 Getting On-line Help

While in StatConcepts, there are two ways you can get help:

1 *The Help menu:* The last menu item at the top of the screen is for getting help. Clicking on this item will open a help window from which you can get general information about using StatConcepts.

2 *Dialog box help:* Every dialog box in StatConcepts has a `Help` button. Clicking on it brings up a help window that briefly describes the procedure for that dialog box and gives instructions on how to fill in the dialog box.

1.3.4 The Data Editor

Clicking on the `Editor` button on the tool bar will open the StataQuest data editor, in which you can enter a new data set or modify the current one (see Figure 1.3). The editor is in standard spreadsheet format. You enter the data one number at a time, using the mouse or the arrow keys and space bar to navigate within the cells of the spreadsheet.

While in the editor, StataQuest will use some default names for the variables created (such as `var1`). To name a variable, double-click at the top of that column and fill in the dialog box.

FIGURE **1.3**
The StataQuest data editor.

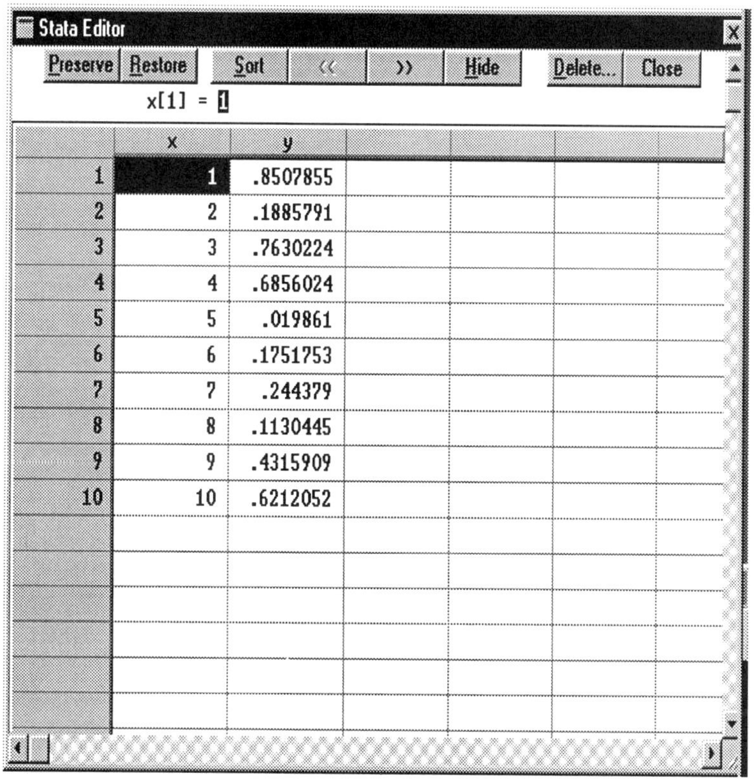
Stata Editor
Preserve
Restore
Sort
<<
>>
Hide
Delete...
Close
x[1] = 1
x
y
1 1 .8507855
2 2 .1885791
3 3 .7630224
4 4 .6856024
5 5 .019861
6 6 .1751753
7 7 .244379
8 8 .1130445
9 9 .4315909
10 10 .6212052

The editor has its own tool bar, which contains several items:

- *Preserve:* This tells StataQuest to keep, or preserve, a copy of the current spreadsheet in its memory.
- *Restore:* This restores the spreadsheet to the way it was the last time you preserved it.
- *Sort:* The data are sorted in the order of the variable where the cursor is located.
- *<<, >>:* This moves the spreadsheet to the left or right.
- *Hide:* This removes the current column from the screen.
- *Delete:* This deletes a variable or case.
- *Close:* This enables you to leave the editor.

1.3.5 The Nine StataQuest Windows

StataQuest has four main windows, as well as five others that can appear during a StataQuest session:

1 *Stata Results:* Textual output from StatConcepts analyses is placed in this window. Figure 1.4 shows the results window for 10 observations in a variable called x containing a random sample from an N(0,1) population. Note that the results window is not scrollable. That is, if there is more text than will fit in the window, some of it will disappear off the top of the window. However, a *log file* will also receive text output, and log files are scrollable (see below). You can copy any part of the current results window to another program by (1) highlighting the part you want using the mouse, (2) using Edit→ Copy, (3) switching to the other program, and 4) using Edit→ Paste in the other program.

2 *Stata Command:* In addition to accessing the various features of StatConcepts by clicking on menu items, you can also access the features by typing commands

FIGURE 1.4
Example of results in the results window.

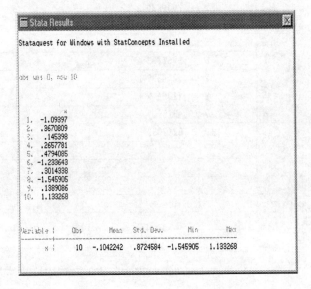

in this window. For example, typing `graph x` will result in a histogram being formed in the graphs window. You can learn about the commands corresponding to any menu item by clicking on `Help` in any dialog box. Note that the command window disappears any time a dialog box is open.

3 *Variables:* A list of variables (and their labels) in the current data set is maintained in this window.

4 *Review:* A list of the commands corresponding to the previous few operations performed in the program is maintained in this window. Double-clicking on any command will reexecute it.

5 *Stata Graph:* Any graphical output is placed in this window. When StatConcepts starts there is no graphical output, and so no graph window. As soon as a graph is produced, the window appears. Any time a new graph is produced, the old one is erased. At any time, you can save or print the current graph window with `File→ Print Graph` or `File→ Save Graph`. You can copy a graph to another program by (1) using `Edit→ Copy Graph`, (2) switching to the other program, and (3) using `File→ Paste` in the other program.

6 *Stata Log:* Clicking on the `Log` button in the tool bar lets you create a file containing all the text output that goes to the results window.

7 *Stata Editor:* Clicking on the `Editor` button in the tool bar will open the StataQuest editor, enabling you to enter or modify a data set.

8 *Stata Browser:* Clicking on the `Browse` button in the tool bar brings up a window similar to the editor window. The current data set will appear in the spreadsheet, but you cannot add or change any of the data.

9 *Stata Help:* Clicking on the `Help` button in the tool bar will open a help window providing instructions on StatConcepts. Similarly, clicking on the `Help` button provided in every dialog box will tell you how to fill in the dialog box.

1.3.6 The Tool Bar

The row of nine buttons below the menu items at the top of the screen is called the *tool bar*. The tool bar contains these features:

1 *Log:* Clicking on this button enables you to either (1) open a log file, (2) temporarily suspend placing text into the log file, or (3) close the log file.

2 *Dialog:* If a dialog box is open but is hidden behind other windows, clicking on this button will bring it to the front.

3 *Results:* If a results window is open but is hidden behind other windows, clicking on this button will bring it to the front.

4 *Graph:* If a graph window is open but is hidden behind other windows, clicking on this button will bring it to the front.

5 *Regraph:* Clicking on this button executes the most recent graph command with the current data.

6 *Editor:* Clicking on this button brings up the StataQuest editor. Also, if the editor is open but hidden behind other windows, clicking on this button will bring it to the front.

7 *Browse:* Clicking on this button brings up the StataQuest browser. Also, if the browser is open but hidden behind other windows, clicking on this button will bring it to the front.

8 *More:* Some menu items in StatConcepts will produce more text output than can all fit in the results window. When this happens, the output will be put in the results window one "screenful" at a time. Clicking on the More button will cause the next screenful of output to be produced.

9 *Break:* As with the More button, clicking on the Break button will stop the display of output.

1.3.7 The Menu Items

There are close to 200 individual menu items in StataQuest and the StatConcepts Labs menu. Those in StatConcepts are documented at great length in the rest of this book; those in StataQuest are outlined in the rest of this chapter. For a more complete description of StataQuest, see *StataQuest 4* by J. T. Anagnoson and R. E. DeLeon, published by Duxbury Press.

The *File* Menu

This menu is typical of file menus in all Microsoft Windows programs. The complete menu is shown in Table 1.1 and contains these items, among others:

1 *New:* This item starts the data editor.
2 *Open:* This item chooses a file in StataQuest format to read into StataQuest.
3 *Save:* This item saves the current data set to a file.

The *Edit* Menu

This menu enables you to communicate with other Windows applications. This menu is also shown in Table 1.1; it contains these items:

1 *Copy or Copy Graph:* If you use the mouse to highlight text in the results window, the *Copy* option will come up. If you click on it, switch to another program, and do Edit→ Paste in that program, the highlighted text will be inserted into that program. If there is an active graphics window in StatConcepts, the *Copy Graph* option will come up. If you click on it, switch to another program, and do Edit→ Paste in that program, the graph will be inserted into that program.

2 *Paste:* It is possible to paste a line of text from some other application into the StataQuest command window, but it is hard to imagine a reason for doing so.

3 *Graph Copy Options:* If you want to copy graphs to another application, this item lets you specify the format (Windows Metafile or Windows Bitmap). It also contains some more advanced options concerning the size and resolution of the graphs you are copying.

TABLE 1.1
The *File, Edit, Prefs,* and *Data* menus and submenus.

File:

| New |
| Open |
| Save |
| Save As |
| Save Graph |
| Print Graph |
| Print Log |
| Import ASCII |
| Export ASCII |
| Exit |

Edit:

| Copy or Copy Graph |
| Paste |
| Graph Copy Options |

Prefs:

Editor	▷
Graph	▷
Results Window	
Default Windowing	
Load Windowing Preferences	
Save Windowing Preferences	

Editor:

| √ Auto-Preserve |

Graph:

| Line Thicknesses |
| Magnifications |
| Colors |
| √ Display Thicknesses on Screen |
| √ Ignore Pen Colors on Print |
| √ Ignore Back Color on Print |
| √ Include Logo on Print |
| Load Graph Preferences |
| Save Graph Preferences |

Data:

Generate/Replace	▷
Label	▷
Edit data	

Generate/Replace:

| Random numbers |
| Sequence |
| Formula |
| Recode |

Label:

| Label dataset |
| Label variable |
| Label values |
| Strings to labels |
| Labels to strings |

The *Prefs* Menu

This menu allows you to configure StatConcepts the way you like it, changing colors, the appearance of graphs, and so on. This menu is also shown in Table 1.1; it contains six items:

1 *Editor:* Suppose you form a data set using the editor, close it, and then open it again to make a change. When you reclose the editor, it will ask you if you really want to make those changes. With the *Editor* item, you can tell StataQuest not to ask this question. This is called *Auto-Preserve*.

2 *Graph:* You can customize your graphs' appearance with nine different items. For example, you use *Line Thicknesses* to specify how thick you want the lines to be, *Colors* to specify the colors you want in the foreground and background in the graphics window, *Save Graph Preferences* after you've specified changes in how things should look, and *Load Graph Preferences* to apply your previous choices. The items *Ignore Pen Colors on Print*, *Ignore Back Color on Print*, and *Include Logo on Print* have to do with the printing of graphs.

3 *Results Window:* This item lets you change the foreground and background colors in the results window.

4 *Default Windowing:* This item restores all of the preferences used by StataQuest the very first time you used it.

5 *Load Windowing Preferences:* This item restores the preferences most recently saved.

6 *Save Windowing Preferences:* This item saves the current preferences.

The *Data* Menu

This menu has three items (see Table 1.1). The second item, *Label*, enables you to label a data set, a variable, or a single data value or to convert a string variable to numerical, or vice versa (for example, convert male-female to 0–1, or vice versa). The third item, *Edit data*, invokes the data editor. The first item, *Generate/Replace*, enables you to generate or replace variables in four ways:

1 *Random numbers:* You can choose from normal, Bernoulli, binomial, chi-squared, exponential, F, t, uniform, and Poisson integers.

2 *Sequence:* You can generate variables consisting of various kinds of sequences (such as $1, 2, \ldots, n$ or $1, 2, 3, 1, 2, 3, \ldots, 1, 2, 3$).

3 *Formula:* You can create variables that are a function of other variables. Constants, arithmetic operations, absolute values, square roots, and the logistic, exponential, and cos/sin functions are available.

4 *Recode:* You can create a new variable consisting of "recoded" values of another variable. This can mean a variety of things, including changing a variable that is all 0's and 1's to all 1's and 2's or converting a variable whose values are between 0 and 1 into a 0–1 variable according to whether the value of the original variable is above or below 0.5.

The *Graphs* Menu

This menu has seven items, all but the last of which has its own menu. Table 1.2 lists the 31 operations that can be performed via the menu; here's an overview of the seven items:

1 *One variable:* From this item, you can get (1) a *Histogram* of a continuous or discrete variable, (2) a *Box plot*, (3) a *Box plot & one-way*, which is a sideways box plot with vertical lines representing each of the data points, (4) a *Stem-and-leaf* plot, (5) a *Dotplot*, and (6) a *Normal quantile plot*.

TABLE 1.2
The Graphs menus and submenus.

Graphs:

One variable	▷
One variable by group	▷
Comparison of variables	▷
Scatterplots	▷
Time series	▷
Quality control	▷
View saved graphs	

One variable:

| Histogram |
| Box plot |
| Box plot & one-way |
| Stem-and-leaf |
| Dotplot |
| Normal quantile plot |

One variable by group:

| Histogram by group |
| Box plots by group |
| Bar charts by group |
| Dotplots by group |

Comparison of variables:

| Box plot comparison |
| Box plot & one-way comparison |
| Bar chart comparison |
| Bar chart comparison by group |

Scatterplots:

| Plot Y vs. X |
| Plot Y vs. X, naming points |
| Plot Y vs. X, with regression line |
| Plot Y vs. X, scaling symbols to Z |
| Plot Y vs. X, by group |
| Scatterplot matrix |

Time series:

| Plot Y vs. X |
| Plot Y vs. obs. no. |
| Plot more than one Y vs. X |
| Plot more than one Y vs. obs. no. |

Quality control:

| Control (C) chart for defects |
| Fraction defective (P) chart |
| Range (R) chart |
| X-bar chart |
| X-bar & R chart |

2 *One variable by group:* From this item, you can get (1) a *Histogram by group*, (2) *Box plots by group*, (3) *Bar charts by group*, and (4) *Dotplots by group*. These items typically are used if you have a variable (called a *data variable*) containing information (such as heights) on a collection of objects (such as people) and another variable (called a *grouping variable*) that classifies each of the objects in the first variable into groups (1 for male and 2 for female, for example). The requested type of plot is produced for the data for each value of the grouping variable (for example, a histogram of heights for males and for females in the same graphics window).

3   *Comparison of variables:* From this item you can get (1) a *Box plot comparison* (see Figure 1.5 for an example), (2) a *Box plot & one-way comparison*, (3) a *Bar chart comparison*, and (4) a *Bar chart comparison by group*. These items differ from those in the *One variable by group* items in that, except for the last operation, there is no grouping variable, but rather two or more variables.

4   *Scatterplots:* From this item, you can get (1) *Plot Y vs. X*; (2) *Plot Y vs. X, naming points*, if you have a string variable with a name for each point (see Figure 1.6 for an example); (3) *Plot Y vs. X, with regression line*; (4) *Plot Y vs. X, scaling symbols to Z*, which is a scatterplot with the size of the plotting symbol proportional to the value of a third variable; (5) *Plot Y vs. X, by group*, which is a set of scatterplots in the graphics window for which there is one point corresponding to each value of a grouping variable (for example, you could get a scatterplot of weight versus height for males and for females if you had all the weights in one variable, all the heights in another, and a third variable of 1 for males and 2 for females); and (6) a *Scatterplot matrix*, which gives a scatterplot for each pair of variables in a set of two or more variables.

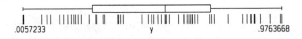

FIGURE **1.5**
Box plot & one-way comparison of two random samples of size 60 from a Uniform(0,1) population. The vertical lines represent the data points.

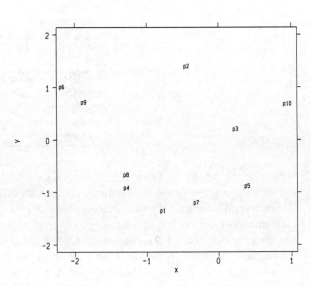

FIGURE **1.6**
Example of a scatterplot with the points named.

5 *Time series:* From this item you can get (1) *Plot Y vs. X*, which connects consecutive points with lines (see Figure 1.7 for an example), (2) *Plot Y vs. obs. no.*, (3) *Plot more than one Y vs. X*, and (4) *Plot more than one Y vs. obs. no.*

6 *Quality control:* From this item, you can get (1) a *Control (C) chart for defects*, when you have one variable containing numbers of defects over time and another with the times; (2) a *Fraction defective (P) chart*, which is the same as the previous operation, except that you have to have a variable with the total number of items at each time; (3) a *Range (R) chart*; (4) an *X-bar chart*; and (5) an *X-bar & R Chart*. These last three items assume that you have samples at a series of times. If you have samples of size 5 at each time, for example, then you must have five variables. The first elements of the five variables are treated as the sample at the first time, the second elements are the second sample, and so on.

7 *View saved graphs:* This item lets you specify a file containing a graph you have previously saved to a file.

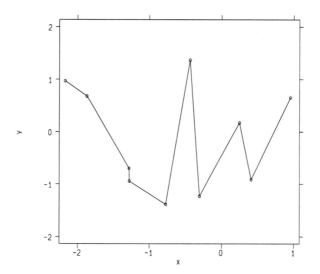

FIGURE **1.7**
Example of a *Plot Y vs. X* time series plot.

The *Summaries* Menu

This menu produces various types of numerical output to the results window for the current data set. This menu is shown in Table 1.3; it contains eight items:

1 *Means and SDs:* This item displays these quantities for one or more selected variables, as well as their minimum and maximum values.

2 *Means and SDs by group:* This item displays these quantities for a data variable and for one or two grouping variables for the *One-way of means* and *Two-way of means* submenu items, respectively.

3 *Median/Percentiles:* This item gives a more detailed numerical summary of a variable, including percentiles, skewness, and so on.

TABLE 1.3
The *Summaries* and *Statistics* menus and submenus.

Summaries:
- Means and SDs
- Means and SDs by group ▷
- Median/Percentiles
- Confidence Intervals
- Tables ▷
- Dataset info
- Describe variables
- List data

Means and standard deviations by group:
- One-way of means
- Two-way of means

Tables:
- One-way (frequency)
- Two-way (cross-tabulation)
- Three-way (by group)

Statistics:
- Parametric tests ▷
- Nonparametric tests ▷
- Correlation ▷
- Simple regression
- Multiple regression
- Robust regression
- Logistic regression
- ANOVA ▷

Nonparametric tests:
- Sign test
- Wilcoxon signed-ranks
- Mann-Whitney
- Kruskal-Wallis
- Kolmogorov-Smirnov

Correlation:
- Pearson (regular)
- Spearman (rank)

Parametric tests:
- 1-sample t test
- 2-sample t test
- Paired t test
- 1-sample test of variance
- 2-sample test of variance
- Normality test
- 1-sample test of proportion
- 2-sample test of proportion

ANOVA:
- One-way nonparametric
- One-way
- Repeated measures
- Two-way
- N-factor ANOVA & ANOCOVA

4. *Confidence Intervals:* This item gives a one-sample t-type confidence interval for the population mean for each selected variable.

5. *Tables:* From this item, you can get three subitems:

   (a) *One-way (frequency):* This produces a list of the distinct values of a variable and the number of times each value occurs (a *frequency table*). There is an option to produce a simple sideways histogram of the frequencies in the results window.

   (b) *Two-way (cross-tabulation):* When given two variables listing the categories that a set of objects belong to according to two classification schemes, this item will produce a two-way table of how many objects belong to each combination of categories (a *cross-tabulation*). Various measures of association and marginal totals and proportions can be obtained.

   (c) *Three-way (by group):* This is like the *Two-way* except that cross-tabulations for two categorical variables are produced for each distinct value of a third categorical variable.

6. *Dataset info:* It is possible to attach a set of notes to a data set describing such things as the origin of the data set, previous results, and so on. This can be

done only through the command window using the *notes* command. Clicking on *Dataset info* results in these notes being displayed in the results window.

7 *Describe variables:* This item gives information on how many variables are defined.

8 *List data:* This item displays in the results window the values of a selected set of variables.

The *Statistics* Menu

This menu, together with the *Calculator* menu, forms the bulk of the statistical inferences in StataQuest. The difference between the two is that for the most part, the *Statistics* items draw inferences from variables while the *Calculator* items use summary statistics (such as sample means and variances) rather than "raw data." Table 1.3 also shows the *Statistics* menu; it contains eight items:

1 *Parametric tests:* This item lets you do the eight statistical tests listed. The items also produce the confidence intervals corresponding to the tests.

2 *Nonparametric tests:* These include two nonparametric versions of the paired t test, an analog of the two-sample t test and of one-way ANOVA, and a test of whether two samples come from populations having the same distribution.

3 *Correlation:* This item will calculate the sample correlation coefficient of two variables and the p value of the test that the correlation is zero, as well as for the rank correlation.

4 *Simple regression:* For user-specified dependent and independent variables, this item will calculate and table all the usual simple regression quantities and then give several graphical and numerical options, including these:

(a) *Plot fitted model*   (b) *Plot residual vs. prediction*
(c) *Plot residual vs. and X*   (d) *Normal quantile plot of residuals*
(e) *Save YHAT as a variable*   (f) *Save residuals as a variable*

5 *Multiple regression:* For one dependent variable and one or more independent variables, this item will (1) perform either a standard multiple linear regression or a stepwise regression, (2) display the usual regression quantities, and then (3) give a set of graphical and numerical options similar to those for simple linear regression. The item can also be used to get the value of the Durbin-Watson statistic for testing residuals for serial correlation.

6 *Robust regression:* This item is similar to the previous one except that it uses a Huber-type robust method for estimating the parameters in the regression model.

7 *Logistic regression:* This item fits the logistic regression model of a binary (0 or 1) dependent variable on a logistic function of a single (binary or continuous) independent variable.

8 *ANOVA:* This item gives a choice of (1) *One-way nonparametric*, which is the Kruskal–Wallis test again; (2) *One-way*, which gives a choice of using either one data variable and one grouping variable or one data variable for each treatment; (3) *Repeated measures*, in which all experimental units are measured under all

treatments; (4) *Two-way*, which uses one data variable containing all observations and then two grouping variables, one specifying the levels of the first factor for each data value and the second specifying the level of the second factor; and (5) *N-factor ANOVA & ANOCOVA*, which analyzes a multifactor experiment in which each experimental unit can have continuous independent variables as well as a response variable. Each of these generates numerical output in the results window and then gives the option of producing some plots such as plots of means with error bars for *One-way* or interaction plots for *Two-way*.

The *Calculator* Menu

This menu and its submenus are listed in Table 1.4. The first 11 items consist of tests and confidence intervals for various situations and require as input only summary statistics. The *Statistical tables* item calculates cumulative areas for specified values of Z, t, χ^2, F, binomial, and Poisson variables, as well as probabilities of single values for binomial or Poisson variables. The *Inverse statistical tables* item calculates percentiles for Z, t, χ^2, and F variables. Finally, two calculators are available: a *Standard calculator* (see Figure 1.8), and an *RPN* (reverse Polish notation) *calculator*.

FIGURE **1.8** StataQuest's *Standard Calculator*.

The *Labs* Menu

This menu is what makes StatConcepts different from StataQuest. The items are listed in Table 1.4 and briefly described in the Preface; in addition, a chapter in this book is devoted to each item.

The *Help* Menu

Clicking on this menu lets you choose which topics from StataQuest to see in a help window.

TABLE 1.4
The *Calculator* and *Labs* menus and submenus.

Calculator:

| 1-sample normal test |
| 2-sample normal test |
| 1-sample t test |
| 2-sample t test |
| 1-sample test of proportion |
| 2-sample test of proportions |
| 1-sample test of variance |
| 2-sample test of variance |
| Confidence interval for mean |
| Binomial confidence interval |
| Poisson confidence interval |
| Statistical tables ▷ |
| Inverse statistical tables ▷ |
| Standard calculator |
| RPN calculator |

Statistical tables:

| Normal |
| Student's t |
| F |
| Chi-squared |
| Binomial |
| Poisson |

Inverse statistical tables:

| Normal |
| Student's t |
| F |
| Chi-squared |

Labs:

| Introduction to Concept Labs |
| Random Sampling |
| Relative Frequency and Probability |
| How Are Populations Distributed? |
| Sampling From 0–1 Populations ▷ |
| Bivariate Descriptive Statistics ▷ |
| Central Limit Theorem |
| Z, t, Chi-square, F ▷ |
| Sampling Distributions |
| Minimum Variance Estimation |
| Interpreting Confidence Intervals |
| Calculating Confidence Intervals |
| Tests of Significance |
| Level of Significance of a Test |
| Calculating Tests of Hypotheses |
| Power of a Test |
| Calculating One-way ANOVA |
| Between and Within Variation |
| Chi-square Goodness of Fit |

Sampling From 0–1 Populations:

| Sampling With & Without Repl. |
| The Negative Binomial Dist. |
| Approx. Binomial Probs. ▷ |

Approximating Binomial Probabilities:

| Poisson Approximation |
| Normal Approximation |

Bivariate Descriptive Statistics:

| Scatterplots I |
| Scatterplots II |
| Least Squares |

Z, t, Chi-square, F:

| Critical Values |
| Normal Curves |
| Chi-square Curves |
| F Curves |
| t Converging to Z |
| Normal Approx. to Binomial |

1.4 Using the Computer As a Statistical Concepts Laboratory

As noted in the Preface, in many ways, the computer is the laboratory for the science of statistics. Most of the ideas of statistics have their origin in phrases like, "If we did this procedure over and over again, then this is what we would see." And again, realistically, the only way to do things over and over again is on a computer. In these labs, we have tried to use graphics to show what, in fact, we would see if we did various things over and over again.

What we do over and over again primarily is to select random samples from a population of numbers. The populations we study in these labs come in two forms:

1. *Finite populations:* Elements of the population can be only one of a finite number of values. One example of this is a *0–1 population*, in which every element can be represented by the number 0 or the number 1. To take random samples from finite populations, we need to know the proportion of elements in the population that have each value.

2. *Continuous populations:* In this kind of population, each element in the population can take on any value in a range of numbers on the real line. For example, people's heights can be any value between some lower limit and some upper limit, although typically heights are rounded to some specified number of decimal places. For this type of population, the proportion of elements having each value is represented by a continuous curve, such as the famous bell-shaped curve. We will see that from this curve we can randomly generate values from the population.

1.4.1 Samples From a Uniform(0,1) Population

Remarkably, statisticians have developed methods for using the computer to get random samples from any kind of population based on the ability to generate a random number in the range from 0 to 1. Suppose we want to obtain n numbers (say, $X_1, X_2, \ldots, X_n$), each taking on any value between 0 and 1, so that all possible values between 0 and 1 are equally likely to occur for each X and no relationship exists between the n X's—that is, knowing the value of any one X tells us nothing about the values of any of the others. The "equally likeliness" is called *uniformity*, and the "no relationship" is called *independence*. Thus, we are looking for n independent uniform random numbers. We denote a population of independent uniform random numbers by Uniform(0,1).

The most common way to use the computer to get independent uniform random numbers is as follows. First, take a large positive integer (which is called a *seed* for reasons that will become apparent) and two other large positive integers (which are called a *multiplier* and a *modulus*, respectively). To illustrate the process, we use multipler $\alpha = 7^5 = 16807$, modulus $M = 2^{31} - 1 = 2147483647$, and seed $X_0 = 1234567$. Next, multiply α times X_0 (which is 20749367569), and find the remainder of the result when it is divided by M (this massive division turns out to be 9 with 1422014746 left over). Then repeat the process with this remainder taking the place of the seed (the next remainder is 456328559).

After doing this n times, we'll have all the n remainders that are between 0 and $M - 1$. If we divide each remainder by M, the resulting n numbers will all be between 0 and 1, as desired. If the multiplier and modulus are chosen carefully (a subject of great research effort in statistics), no number will actually equal 0, which is important because if any number equaled 0, all subsequent numbers would also be 0. Note why the first number in the sequence is called a seed, namely, all the others are generated from it. StatConcepts maintains its own seed for all random number generation. You can specify your own seed in the *Introduction to StatConcepts* Lab.

Clearly, these numbers are not "random" in the true sense. However, it turns out that if we use good multipliers and moduli, almost any statistical test for uniformity and independence of a set of numbers will find the numbers to be indistinguishable from truly random numbers.

1.4.2 Random Integer From 1 to N

If we have a random number in the interval from 0 to 1 (which can't be exactly 0 or exactly 1) and we multiply it times N and throw away the decimal part of the result, then the result will be an integer that has equal chance of being $0, 1, \ldots, N-1$. Thus, if we add 1 to the integer, we have a random integer from 1 to N.

1.4.3 Sampling With Replacement From a Finite Population

If we number the N elements in a finite population, then we can repeatedly randomly choose elements by generating random integers from 1 to N and selecting the corresponding elements of the population.

1.4.4 Sampling Without Replacement From a Finite Population

This situation is different from the previous one because once we have selected an element from the population, we have to make sure it doesn't get selected again. The most common way to do this is as follows. First, write down a list of the integers from 1 to N. Next, generate a random integer from 1 to N and exchange the number in the list whose index is the generated random integer with the last element in the list. Now generate a random integer from 1 to $N-1$ and exchange the number in that place in the new list with the $(N-1)$st element. Do this n times, each time reducing the upper limit on the random integer by 1. The last n elements in the final list will be the numbers of the elements in the population in the random sample of size n.

For example, to do the Lotto example in the *Relative Frequency and Probability* Lab, we have to randomly choose (without replacement) six of the numbers from 1 to 50. We start by visualizing balls numbered 1 through 50 sitting in slots numbered 1 through 50. Then we generate a random number from 1 to 50 and exchange the ball in the 50th slot with the one in the slot whose number (not ball) is the random number. Now we get a random number from 1 to 49 and exchange the ball in that slot with the one in the 49th slot. After we do this four more times, the balls in the last six slots are the random sample.

1.4.5 Sampling From a Continuous Population

Another remarkable fact is that statisticians have shown that for any continuous population, a function exists such that when it is applied to a random number in the interval from 0 to 1, the result is a randomly selected element from the desired continuous population. For example, to get a random number from an exponential population having mean 1 (see the *How Are Populations Distributed?* Lab), we need only calculate the negative of the logarithm of a randomly chosen element from a Uniform(0,1) population. Similar functions exist for any distribution.

1.5 The *Introduction to Concept Labs* Lab

If you select Introduction to Concept Labs from the Labs menu, the dialog box shown in Figure 1.9 appears on the screen. It has the following box and three buttons:

1. seed: As discussed in the previous section, the random number generator used by StatConcepts requires the use of a seed. Every time StatConcepts starts, it will use the value 123456789 as the seed. Then, every time it uses a random number generator, it keeps the last remainder to be used as the seed for the next time it generates random numbers. If you want to specify your own seed, you can enter it in this box. The value should be a large positive integer. Note that seeds are important for another reason. If you set your own seed, perform an experiment, and then reset the same seed and repeat the experiment, the results will be the same each time. This makes it possible to reproduce the same results during different sessions.
2. Run: Clicking on this button sets the seed specified in the seed box.
3. Help: Clicking on this button brings up a help window that briefly describes each of the labs in the collection.
4. Close: Clicking on this button closes the lab and returns you to the StatConcepts menus.

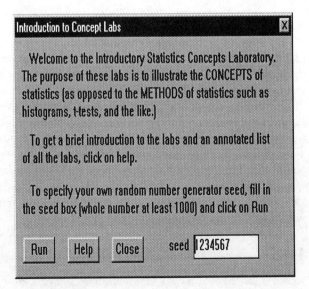

FIGURE **1.9**
Dialog box for the *Introduction to Concept Labs* Lab.

Random Sampling

In many fields, information is required about some large collection of objects or individuals. In most cases, the collection is so large that it is impossible or impractical to observe each individual or object. Sampling from that population allows us to describe characteristics of the population without having to involve every individual.

2.1 Introduction

In this lab, we will illustrate the meaning and implications of choosing a sample randomly. As we progress, we will examine the behavior of the values of the sample mean for many different samples of size n from the same population. We will also study how this behavior depends on the number of elements in the samples being studied.

2.1.1 Some Basic Ideas

Before beginning the lab, an overview of some fundamental concepts of random sampling and distributions will be helpful.

1 Suppose information is required about some large collection of individuals called the *population*. The number of individuals in the population is typically denoted by N. In most cases, N is so large that the entire population cannot be studied. Therefore, before any information can be obtained, a *sample* must be chosen.

2. If samples are not properly chosen, the results obtained may be misleading. Randomly choosing individuals to be included in a sample helps ensure that no bias is introduced.

3. A common misconception is that a truly random sample will consist of individuals evenly dispersed throughout the population. However, the proper definition of a *(simple) random sample* of size *n* is that it is chosen in such a way that every possible set of *n* individuals in the population has an equal probability of being selected.

4. Each individual in the population has a measureable value of some property (for example, the thickness of a semiconductor chip or the color of a person's eyes). A plot of how often each measurement occurs is called the *distribution of the population*.

5. Numerical characteristics for the population as a whole (for example, the average thickness of all semiconductor chips or the percentage of people with blue eyes) are called *parameters*.

6. After a random sample is chosen, a *statistic* that estimates the parameter is computed. The observed value of the statistic depends entirely upon which individuals are included in the sample. Because different samples will result in different values of the statistic, all possible values of the statistic also follow a pattern, called the *sampling distribution of the statistic*.

2.2 Objectives

1. Illustrate random sampling.
2. Introduce the concept of the distribution of a population.
3. Introduce the sampling distribution of a statistic.
4. Illustrate the effect of sample size on the sampling distribution of the sample mean.

2.3 Description

Start the lab by clicking on `Random Sampling` in the `Labs` menu.

1. Ten columns containing 10 boxes, labeled 1 through 100, will appear (see Figure 2.1). The collection of boxes is the population, with each of the boxes an individual in the population.

2. Each individual in the population has a measurable value defined by the property to be studied. The number on the horizontal axis is the value for the individuals in that column. For example, each box in the first column of boxes has the value 0.05, each box in the second column has the value 0.15, and so on.

3. Five of the 100 boxes are randomly selected and colored red.

FIGURE 2.1
Example of the initial *Random Sampling* graph.

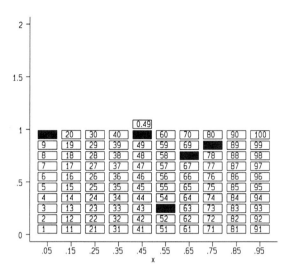

4 The sample mean of the measured value of the property of the five boxes is calculated. A yellow box containing that value appears on top of the column whose value is closest to the value of the sample mean.

5 A dialog box containing these options appears (see Figure 2.2):

   (a) Reset: Clicking on this button restarts the lab.

   (b) Sample: Clicking on this button causes the lab to select a new sample and calculate its mean. The boxes included in the previous sample are redrawn in blue, and the boxes in the new sample appear in red. The yellow box containing the sample mean from the previous sample remains. A new yellow box containing the calculated value of the sample mean for the new sample appears over the column with the value closest to the new sample mean. This process may be repeated as many times as needed.

FIGURE 2.2
Dialog box for the *Random Sampling* Lab.

(c) Help: Clicking on this button opens a help window containing information about the lab.

(d) Close: Clicking on this button closes the lab and returns you to the StatConcepts menus.

(e) n: To run the lab with a sample size other than $n = 5$, type in the desired value for n. Changing the sample size restarts the lab and discards any previously accumulated information. Note that n must be at least 1 and at most 20.

2.4 Guided Tour of the Lab

Because there are only 10 possible values of the elements in the population 0.05, 0.15, ..., 0.95, the distribution of the population is *discrete*. Because each of the 10 possible values occurs in the population with the same relative frequency (or probability), the distribution of the population is called a *uniform distribution*. In the *How Are Populations Distributed?* Lab (Chapter 4), many population distributions will be examined, including a continuous uniform distribution.

2.4.1 What Does Randomness Look Like?

Each time you click the Sample button, you will get a new random sample from the population of 100 boxes. Each set will be totally unrelated to any previous sample; that is, the samples are *independent*.

Now you should repeatedly click the Sample button and pay close attention to where the red boxes appear. Do they seem to be spread out evenly among the 100 boxes, or do you see cases in which there are surprisingly many red boxes in the same row or column of boxes? You should, in fact, see such seemingly nonrandom samples surprisingly often. Rest assured, however, that the samples are being drawn randomly; that is, each time a sample of five boxes is drawn, every possible set of five boxes has the same chance of being selected.

If you get 25 such samples, you will see a graph similar to the one shown in Figure 2.3. Each person doing the lab will get a slightly different graph because the lab is randomly selecting the samples.

2.4.2 The Behavior of Sample Means

Another thing to be aware of when you get a new sample is where the yellow box representing the average of the values of the five boxes appears for that sample. Notice that most samples have a sample mean near the middle of the columns of boxes. Why is this? For the average of the values of five boxes to be near the ends of the columns, that is, close to 0.05 or 0.95, all five of the boxes would have to be around 0.05 or around 0.95, and this is very unlikely.

This effect gets even stronger if we have more than five boxes in the samples—that is, the means for the samples will tend to cluster even closer to the middle of the columns of boxes. To illustrate this, change the sample size to $n = 10$ and take at least 25 samples. You should get a graph similar to the one shown in Figure 2.4,

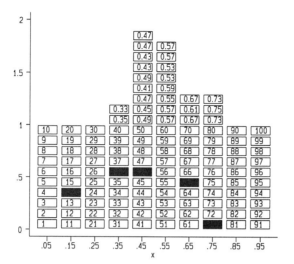

FIGURE 2.3
Example of *Random Sampling* after generating 25 samples for n = 5.

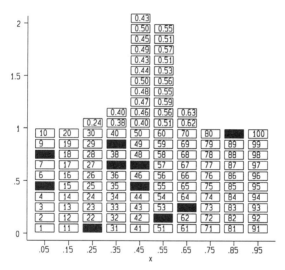

FIGURE 2.4
Example of *Random Sampling* after generating 25 samples for n = 10.

where only a few of the yellow boxes are not above the middle two columns of boxes. In doing this for samples of size 20, it is not unusual to fill up these middle two columns of boxes before getting a yellow box in any other column.

2.5 Summary

We discussed three important ideas in this chapter:

1. Randomness does not mean evenly spread out, and it is possible to obtain samples that don't look evenly spread out.
2. Regardless of the sample size, the sample means are clustered around the center of the population distribution.

3. As the sample size increases, the sample means fall closer to the center of the population distribution more often; that is, as the sample size is increasing, the variability of the sample means is decreasing.

2.6 Lab Exercises

For each of $n = 3$, $n = 9$, and $n = 15$, generate 30 samples.

1. For each sample size, count the number of samples having two or more red boxes in the same row or column.
2. Print final graphs for each value of n. For each of the graphs, answer the following questions:

    (a) Around what value are the sample means centered?

    (b) What is the difference between the smallest and largest values for the sample means? How does this difference behave as the sample size increases? For which sample size does the distribution of the means appear wider?

3

Relative Frequency and Probability

An extremely popular example of random sampling without replacement exists and is used around the world many times, every day—the lottery!

3.1 Introduction

This lab replicates the game in which 50 balls, labeled 1 through 50, are placed in a hopper that continually mixes the balls. One ball at a time is drawn and placed on a stand so that its label is visible. This process is repeated until a random sample (without replacement) of six balls has been chosen. Before these balls are selected, players choose six numbers between 1 and 50. If the numbers a player chooses match the numbers on the balls removed from the hopper, the player wins the jackpot!

3.1.1 Some Basic Ideas

Players of a popular lottery game in Texas were suspicious that the winning numbers were not being randomly chosen because a seemingly disproportionately high number of winning combinations included two or more balls with consecutive numbers (for example, a set of winning numbers containing the balls labeled 15 and 16). These people argued that the selected numbers were not "spread out" enough to be random. A Texas A&M statistician was employed to determine whether the samples were indeed random. By using the concepts we will study in this lab, along with more sophisticated statistical techniques, he concluded that the samples were randomly selected. In this lab, we will see exactly what it means for a sample to be randomly selected and how the statistician's conclusion is easily justified.

A sample is randomly selected if it is chosen in such a manner that *all possible subsets of size n in the population have the same chance of being selected*. This includes samples with two or more balls with successive numbers.

3.2 Objectives

1. Illustrate random sampling.
2. Develop the idea of the probability of an event occurring as the long-run relative frequency of the event.

3.3 Description

To begin the lab, select Relative Frequency and Probability from the Labs menu. This opens the dialog box shown in Figure 3.1 and then a graphics window containing a picture like the one in Figure 3.2. In the picture you see on the screen, 6 balls will be drawn in yellow, and the remaining 44 in blue.

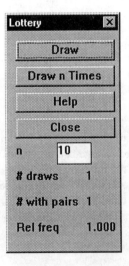

FIGURE 3.1
Dialog box for the *Relative Frequency and Probability* Lab.

To the immediate right of the dialog box is a graphics window containing a collection of 50 balls, labeled 1 through 50, aligned from left to right in five rows of 10 balls each. The heading # draws 1 provides information on how many samples of 6 balls have been randomly selected for this session. The number (which is 1 when the lab is started) will be incremented by 1 each time a new sample is drawn. The balls highlighted in yellow are the balls that were randomly selected for the first sample.

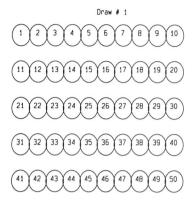

FIGURE 3.2
Example of *Relative Frequency and Probability* Lab.

The dialog box has four buttons at the top:

1. `Draw`: Clicking on this button causes the lab to draw a new random sample of six balls. The balls in the previous sample are redrawn in blue, and the balls selected for the new sample are drawn in yellow. The value for `# draws` (in both the dialog box and the graph heading) is incremented by 1. If the new sample contains a pair of balls that are adjacent to each other (for example, 23, 24), the `# with pairs` is also updated by 1; otherwise, this remains the same. The `Rel freq` is also updated by dividing the (possibly new) value for `# with pairs` by the new value for `# draws`. You can click on the `Draw` button as many times as you wish.

2. `Draw n Times`: Clicking on this button causes the lab to draw n new random samples of six balls, where n is the value specified in the box to the right of n. Each time a sample is chosen, the balls in the previous sample are redrawn in blue, and the balls selected for the new sample are drawn in yellow. The initial value of n is 10. In this case, clicking on `Draw n Times` will result in 10 random samples of six balls each. For each new sample drawn, the value for `# draws` will be increased by 1. Each time a sample contains a pair of balls that are adjacent to each other, `# with pairs` is increased by 1. For each new sample, the value of `Rel freq` is also updated by dividing the (possibly new) value for `# with pairs` by the new value for `# draws` for that sample. You can click on this button as many times as you wish.

3. `Help`: Clicking on this button opens a help window containing information about the lab.

4. `Close`: Clicking on this button closes the lab and returns you to the StatConcepts menus.

Under these buttons is a smaller box where you can type in the number of draws used in `Draw n Times`. Beneath this are three lines containing this information:

1. `# draws`: Every time a new sample is chosen, the number to the right of `# draws` is increased by 1. It will continue to be incremented by 1 until you close the lab.

2 `# with pairs`: If the new sample drawn contains a pair of balls that are adjacent to each other (for example, balls 10 and 11), the number to the right of `# with pairs` is incremented by 1. Otherwise, the number remains the same.

3 `Rel freq`: The number to the right of `Rel freq` is the ratio of the `# with pairs` to the `# draws`. It tells us the proportion of samples taken thus far that had a pair of adjacent balls. For every new sample that is drawn, this number is updated by taking the quotient of the values for `# with pairs` and `# draws` for that sample.

3.4 Guided Tour of the Lab

In this guided tour, we will see that randomly choosing a sample of size n means choosing the sample such that every subset of size n from the population has an equal chance of being selected. We will develop the idea that the *probability of an event occurring* is the long-run relative frequency the event occurs.

When you run the lab, notice the first graph that appears. In the sample of six balls, do you have two or more adjacent balls (like a 46 and 47) in the sample? You may or may not. If you do, the `# with pairs` and `# draws` will both be 1. If not, the `# with pairs` will be 0. If you are working with a group of people in a computer lab, some graphs will have two or more adjacent balls, and some will not. This is because every sample that is chosen is independent of every other sample.

Now click on `Draw`. Are there two or more adjacent balls in this sample? Continue to click on `Draw` until a sample is selected that contains at least two adjacent balls. How many draws were necessary? You should get such a sample relatively early in the resampling process.

Next, click on `Draw n Times`. What is the value for `# with pairs`? How many samples have you chosen? If you divide `# with pairs` by `# draws`, you should get `Rel freq`. For example, notice in Figure 3.3 that $7/14 = 0.500$.

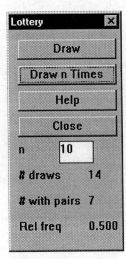

FIGURE 3.3
Dialog box after
`# draws = 14`.

Now change the value of n to 100, and click on Draw n Times. How many samples have pairs? What is the relative frequency for the number of pairs that occur?

Although it is not easy, it can be shown that the proportion of all possible samples of size 6 drawn without replacement from a population of size 50 that do *not* contain a pair is $\binom{45}{6}/\binom{50}{6} = 8145060/15890700 = 0.5126$. Therefore, if we continually chose samples of size 6, we would eventually reach a relative frequency for the number of pairs of $1 - 0.5126 = 0.4874$. Technically speaking, slightly fewer than half (48.74%) of the samples that could be chosen contain pairs.

To end the tour, determine how many more samples are necessary so that a total of 200 draws will be taken. Change the value of n to that value. Based on what we have just seen, you should expect approximately $(200)(0.4874) = 97.47$ (97 or 98) samples containing pairs. Click on Draw n Times. How many samples from the 200 contained pairs? Our experiment resulted in # with pairs being 90 (see Figure 3.4). You probably did not get 90, but that's okay. Because every sample was randomly selected, we have no assurances that there will be exactly 97 or 98 samples with pairs. This is just a guideline so we will know what kind of behavior to expect. If we select another 200 samples, we are not likely to get 90 samples with pairs again. We also will probably not get 97 or 98. (Try it and see!) But we are likely to get some value for # with pairs close to 97 or 98. The important thing to remember is that if we could choose all possible samples of size 6 from this population, or if we could continue sampling forever, the resulting relative frequency in either case would be 0.4874.

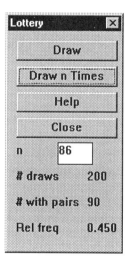

FIGURE 3.4
Dialog box after # draws = 200.

One final note: Realistically, we take only one sample. However, given the information we have seen in this lab, we should not be surprised when a sample contains a pair. Likewise, lottery players should not be at all surprised to see winning numbers with at least two consecutive values.

3.5 Summary

In this lab we have learned several things:

1. A random sample of size n is properly defined as a sample chosen so that all possible subsets of size n in the population have an equal probability of being chosen.

2. This fact does not guarantee the individuals chosen for a particular sample will be spread out. In fact, we have actually observed for a sample of size 6, drawn without replacement from a population of size 50, that samples containing pairs would be selected a little less than half the time.

3. The relative frequency of pairs occurring in a sample tells us the proportion of samples drawn that contained pairs. In our special case, this long-run relative frequency is 0.4874, which is exactly the probability of selecting a sample with a pair; that is, the proportion of all samples of size 6, drawn without replacement from a population of size 50, that contain a pair is 0.4874.

3.6 Lab Exercises

1. Select $n = 10$ draws. How many samples do you expect to get with pairs?
2. Select samples until you obtain five samples that have pairs. How many samples did you need to select? What is the relative frequency of the number of pairs?
3. Select $n = 150$ samples. How many samples do you expect to have pairs? How many actually did have pairs? What is the relative frequency of the number of pairs?
4. Select $n = 200$ draws five times so that you have a total of 1000 draws. How many samples do you expect to have pairs? What is the relative frequency of the number of pairs?

4

How Are Populations Distributed?

A crucial idea in statistics is the distribution of a population, that is, a graph of the values in the population and the likelihood of each value occurring.

4.1 Introduction

4.1.1 Some Basic Ideas

1. If we plotted a histogram of an entire population using many bars and then passed a smooth curve through the tops of the bars, we would get what is called the *distribution curve* of that population. Statisticians have found a wide variety of mathematical functions that when plotted look like the distribution curve of populations. A simple example of this is the bell-shaped curve (the normal distribution). If we were to plot the function

$$f(x) = \frac{1}{\sqrt{2\pi}\,\sigma} e^{-(x-\mu)^2/2\sigma^2},$$

   we would get a curve that looks just like a bell-shaped curve, which, in turn, would look like the histograms of many different populations (for example, height, weight, IQ, and so on).

2. Distributions come in all shapes and sizes. This lab looks at 14 different kinds of distributions.

3. Not all populations have a bell-shaped distribution. In this lab, we will see many distributions that look nothing like the normal curve.

4 Distributions come in families (such as the normal family). Each curve in a family is determined by the values of a set of *parameters*. In the formula above, if we graphed $f(x)$ first for $\mu = 0$ and $\sigma = 1$ and then for $\mu = 1$ and $\sigma = 1$, we would get a different curve. Both curves would be bell-shaped, but the second one would be shifted slightly to the right (because it turns out that μ represents the mean of the population). The values for μ and σ are examples of parameters. (Figure 4.2 later in the chapter shows graphs of the bell-shaped curve for five different sets of parameters; in the figure, the numbers in parentheses represent μ and σ^2.)

5 Different populations (people's height and lengths of lives, for example) will have distributions in different families (for example, heights in the normal family, lifetimes in the exponential or Weibull families).

6 Sometimes, the parameters have a special meaning with regard to the population being studied (for example, the parameters of a normal curve are the mean and variance of the population).

7 Determining which family of distributions a population's distribution falls into is called *modeling*. Picking out which member of that family best fits the histogram of a sample from that population is called *parameter estimation*.

8 If we know the family of distributions and which member of that family best models a population, then we can make numerical statements about that population (for example, what its median or percentiles are).

9 Much of modern science consists of comparing the distributions of two or more populations. In many cases, it is reasonable to assume that the distributions will all be in the same family of curves. Therefore, the scientific task is simply to compare the parameters of the different populations, because if the distributions are all in the same family, then the curves describing the populations will all be identical. This is what the science of *inferential statistics* or *parametric inference* is about: the study of procedures for comparing the parameters of population distributions.

4.2 Objectives

This lab will illustrate each of the basic ideas just outlined.

4.3 Description

To begin the lab, select How Are Populations Distributed? from the Labs menu. This opens the dialog box shown in Figure 4.1, which describes the basic ideas of the lab and contains two items labeled Distribution Family and Option, as well as the usual buttons labeled Run, Close, and Help.

There are two different ways the lab can operate depending on which of two options you choose in the Option box:

1 Curves: If you select Curves and then select a distribution in the Distribution Family item and click on Run, the lab will draw several

FIGURE 4.1
Dialog box for the *Distributions* Concept Lab.

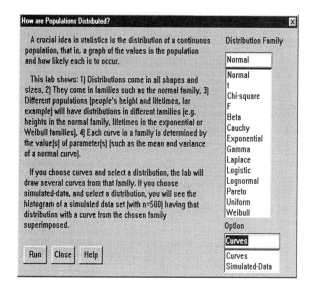

distribution curves from that family. Next to each curve are listed the values of its parameters according to the formulas at the end of this chapter.

2 `Simulated-Data`: If you select `Simulated-Data` and then select a distribution in the `Distribution Family` item and click on Run, the lab will draw the histogram of an artificial, computer-generated data set (also called a "simulated" data set) from a population having a theoretical distribution curve from the specified family of curves. The theoretical curve is superimposed on the histogram. Each time you click on Run for a particular family, you will get another sample from that population.

The three buttons are used as follows:

1 `Run`: Clicking on this button executes the lab using the values specified in the boxes in the dialog box.
2 `Close`: Clicking on this button closes the lab and returns you to the StatConcepts menus.
3 `Help`: Clicking on this button opens a help window containing information about the lab.

4.4 Guided Tour of the Lab

In this guided tour, we visit each of the families listed in the `Distribution Family` box.

4.4.1 The Normal Family

In the `Distribution Family` item, click on `Normal`, and in the `Option` item, click on `Curves`. Then click on `Run`. You will see the graph shown at the top of

Figure 4.2. Here are five members of the famous family of bell-shaped curves, which have the following characteristics:

1. The normal family is used to model a wide variety of measurements such as heights, weights, IQ, and so on.
2. They are also used to describe the *sampling distribution of sample means*. Thus, visualize having a population with parameters denoted by mean μ and variance σ^2, selecting millions of random samples (all of the same size n) from the population, and calculating the sample mean of each sample. If we drew a histogram of this population of $\bar{X}$'s, it would look very much like one of the normal curves—namely, the one having the same mean as the original or *parent* population and variance equal to σ^2/n (that is, the variance of the parent population divided by the size of the samples being used to calculate $\bar{X}$). This is called the *Central Limit Theorem*.

FIGURE **4.2**
Members of the normal family and a simulated data set.

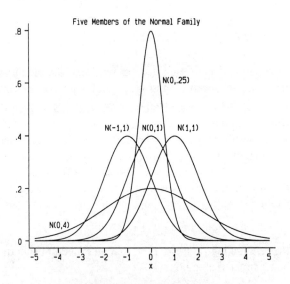
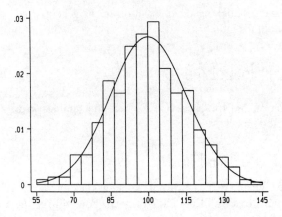

3. The parameters of the family (the numbers in parentheses) are the mean and variance of the population.

4. Looking at the three curves having variance 1 shows that changing the mean from -1 to 0 to 1 changes only the *location* of the curve, not the *scale* of the curve.

5. On the other hand, the three curves having mean 0 show that the variance determines how spread out the curve is; the larger the variance, the greater the spread. Because the total area under any one curve is 1, the more spread out a curve is, the "shorter" it must be.

To study these ideas further, see the *Central Limit Theorem* Lab (Chapter 7) and the *Normal Curves* Lab under the Z, t, χ^2, and F Lab menu (Chapter 8).

Now click on Simulated-Data in the Option item and then click on Run. You will see a histogram and distribution curve similar to the one shown at the bottom of Figure 4.2. The histogram is for a random sample of 500 observations from the normal curve having mean 100 and variance 225. If you click on Run again, you'll get the histogram of another sample.

4.4.2 The Student *t* Family

In the Distribution Family item, click on t, and in the Option item, click on Curves. Then click on Run. You will see the graph shown at the top of Figure 4.3. This family is not typically used to describe populations observed in nature, but rather to describe what would happen if we selected millions of random samples from a normal population having mean μ and then calculated

$$t = \frac{\bar{X} - \mu}{s/\sqrt{n}}$$

for each sample. If we did this and drew a histogram of all the resulting values of t, it would look very much like one of the curves shown in Figure 4.3. Looking at the middle of the curves, the bottom one is the curve we would get if we took samples of size $n = 2$, the next one would be for samples of size $n = 3$, and so on. For technical reasons, the parameter of each curve is actually $n - 1$, not n. The parameter is called the *degrees of freedom*.

The top curve in the figure is actually the Z curve, that is, the normal curve for mean 0 and variance 1. Thus, this figure illustrates the fact that as degrees of freedom increase, the t curves get closer and closer to the Z curve.

To study these ideas further, see the *t Converging to Z* Lab under the Z, t, χ^2, and F Lab menu (Chapter 8) and the *Sampling Distributions* Lab (Chapter 9).

Now click on Simulated-Data in the Option item and then click on Run. You will see a histogram and distribution curve similar to the one shown at the bottom of Figure 4.3. The histogram is for a random sample of size 500 from a t population having 5 degrees of freedom. This corresponds to calculating the t statistics above for 500 samples of size 6 (degrees of freedom are one less than the sample size) from a normal population. Clicking on Run repeatedly will generate a new sample of t values. You should do this a few times and note how well the theoretical curve matches the histogram of the 500 values.

FIGURE **4.3**
Members of the *t* family and a simulated data set.

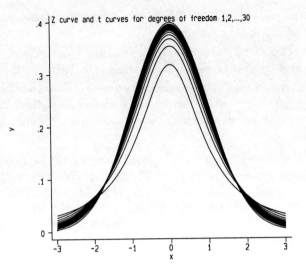

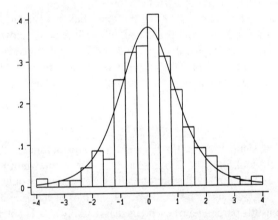

4.4.3 The Chi-square Family

In the Distribution Family item, click on Chi-square, and in the Option item, click on Curves. Then click on Run. You will see the graph shown at the top of Figure 4.4. Like the *t* family of curves, the χ^2 family is not typically used to describe populations observed in nature, but rather to describe what happens if we select millions of random samples from a normal population having variance σ^2 and calculate

$$\chi^2 = \frac{(n-1)s^2}{\sigma^2}$$

for each sample. If we did this and drew a histogram of all the resulting values of χ^2, it would look very much like one of the curves shown at the top of Figure 4.4. The curve on the left is the one we would get if we took samples of size $n = 11$, the next

FIGURE 4.4
Members of the chi-square family and a simulated data set.

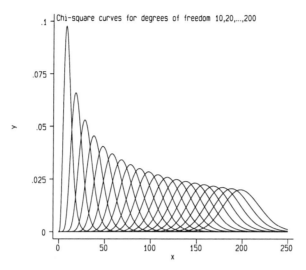

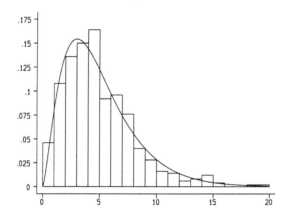

one would be for samples of size $n = 21$, and so on. Again, the parameter of each curve is called degrees of freedom and is one less than the size of the samples being taken in calculating the χ^2 values.

Notice that there are no negative values on the horizontal axis, because the formula for χ^2 shows that it can't be negative. Also, as degrees of freedom increase, the curve moves to the right (the mean of the population increases) but gets shorter and more spread out (the variance also increases), and the curve becomes more and more symmetrical and centered on the value of the degrees of freedom. This is because as the sample size increases, the sample variance should get closer to the population variance. Therefore, their ratio will get close to 1, and the value of χ^2 (which is the value of the degrees of freedom times the ratio of sample to population variance) will get closer to the value of the degrees of freedom.

To study these ideas further, see the χ^2 *Curves* Lab under the *Z, t, χ^2, and F Lab* menu (Chapter 8) and the *Sampling Distributions* Lab (Chapter 9).

Now click on Simulated-Data in the Option item and then click on Run. You will see a histogram and distribution curve similar to the one shown at the bottom of Figure 4.4. The histogram is for a random sample of size 500 from a χ^2 population having 5 degrees of freedom. This corresponds to calculating the χ^2 statistics above for 500 samples of size 6 from a normal population. Clicking on Run repeatedly will generate a new sample of χ^2 values. You should do this a few times and note how well the theoretical curve matches the histogram of the 500 values.

4.4.4 The F Family

In the Distribution Family item, click on F, and in the Option item, click on Curves. Then click on Run. You will see the graph shown at the top of Figure 4.5. Like the t and χ^2 families of curves, the F family is used to describe what happens if we take millions of random samples and calculate something. In this case, we actually take millions of *pairs* of samples. We visualize having two normal populations, the first having variance denoted by σ_1^2 and the second by σ_2^2. Then we repeatedly take a sample of one size (call it n_1) from the first population and a sample of size n_2 (which could be the same as n_1) from the second population and calculate the F statistic

$$F = \frac{s_1^2/\sigma_1^2}{s_2^2/\sigma_2^2},$$

where s_1^2 is the variance of the first sample and s_2^2 is the variance of the second sample. Note that we have to get the two samples in each pair of samples independently. That is, knowing the values in the first sample should not have any influence on the values in the second sample.

If we obtained millions of values of F and drew a histogram of them, it would look very much like one of the curves shown in Figure 4.5. The lowest (and widest) curve is the one we would get if we took samples of size $n_1 = n_2 = 11$, the next higher one is for samples of size $n_1 = n_2 = 21$, and so on. Because for each F value that we calculate we have two samples, the resulting curves have two parameters, namely, the *numerator* and *denominator* degrees of freedom. Each of these is again one less than its corresponding sample size. They are called numerator and denominator degrees of freedom because they correspond to samples in the numerator and denominator of the formula for F.

Again, there are no negative values on the horizontal axis. As degrees of freedom increase, the curves become taller and narrower (the variance decreases), they get more symmetrical, and they become centered on the value 1. This is because as sample sizes increase, the sample variances will get closer to the population variances. Therefore, the numerator and denominator of F will each be close to 1, as will the whole fraction.

To study these ideas further, see the *F Curves* Lab under the Z, t, χ^2, and F Lab menu (Chapter 8) and the *Sampling Distributions* Lab (Chapter 9).

Now click on Simulated-Data in the Option item and then click on Run. You will see a histogram and distribution curve similar to the one shown at the bottom of Figure 4.5. The histogram is for a random sample of size 500 from an F population having 10 (numerator) and 15 (denominator) degrees of freedom. This corresponds

FIGURE 4.5
Members of the *F* family and a simulated data set.

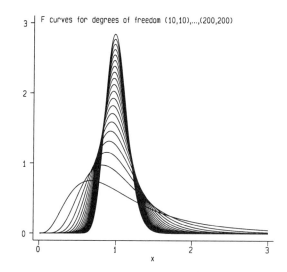

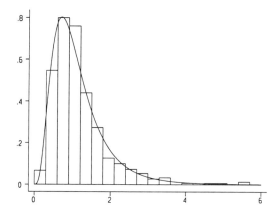

to calculating the *F* statistics above for 500 pairs of samples of size 11 and 16 from two normal populations. Clicking on Run repeatedly will generate a new sample of *F* values. You should do this a few times and note how well the theoretical curve matches the histogram of the 500 values.

4.4.5 The Beta Family

In the Distribution Family item, click on Beta, and in the Option item, click on Curves. Then click on Run. You will see the graph shown at the top of Figure 4.6. This family is used primarily in statistical theory and not very often for modeling data from populations. We include it because it illustrates that a single family of distributions can include members of widely varying shapes. The curves have two parameters, p and q. Unlike the normal family, these parameters are not the mean and variance of the population. Rather, if you look at the formulas in the table

FIGURE 4.6
Members of the beta family and a simulated data set.

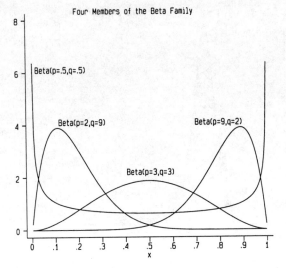

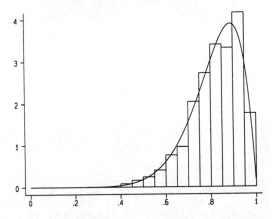

at the end of this chapter, you will see that the mean is $p/(p+q)$ while the variance is a more complicated function of p and q. For example, the mean for $p = 9$ and $q = 2$ is $9/(9+2) = 9/11 = 0.8182$.

The beta family is used for populations whose members must be between 0 and 1.

Now click on Simulated-Data in the Option item and then click on Run. You will see a histogram and distribution curve similar to the one shown at the bottom of Figure 4.6. The histogram is for a sample of size 500 from the beta curve having $p = 9$ and $q = 2$. You should click on Run a few times to see how well the curve matches the histogram.

4.4.6 The Cauchy Family

In the Distribution Family item, click on Cauchy, and in the Option item, click on Curves. Then click on Run. You will see the graph shown at the top of

Figure 4.7. The plot shows one member of the Cauchy family as well as the standard normal curve. The Cauchy family is similar to the normal family except that its members have much heavier *tails* than the normal; that is, values in a population away from the middle are more likely to occur. Thus, the Cauchy family is often used to model data having outliers, that is, data having a few values that are very different from the rest.

Now click on Simulated-Data in the Option item and then click on Run. You will see a histogram and distribution curve similar to the one shown at the bottom of Figure 4.7. The histogram is for a sample of 500 elements from a population having the standard Cauchy distribution. It illustrates the heavy-tailedness of the Cauchy. Note the boxes in the histogram at each end of the axis. Any observation outside the range of the axis is included in these outside boxes, which could be interpreted as outliers. Click on Run several times to see how many outlying values you get from one sample to another.

FIGURE **4.7**
A member of the Cauchy family and a simulated data set.

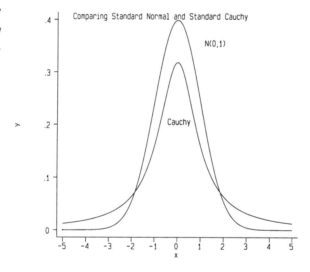

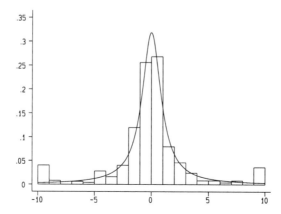

4.4.7 The Exponential Family

In the Distribution Family item, click on Exponential, and in the Option item, click on Curves. Then click on Run. You will see the graph shown at the top of Figure 4.8. Here, we have four curves, for populations having means 2, 3, 4, and 5, respectively. All the values on the horizontal axis are positive, so these curves can be used only for variables that are positive. Further, for each curve, values on the horizontal axis get less and less likely as the value in the population increases. Because of these characteristics, curves in the family of exponential distributions are used to describe populations of lifetimes (for example, light bulbs, people, and so on). They are also used to describe times between certain kinds of events (for example, times between failures of a machine).

Now click on Simulated-Data in the Option item and then click on Run. You will see a histogram and distribution curve similar to the one shown at the bottom

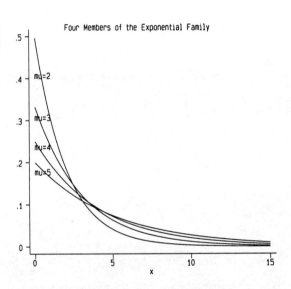
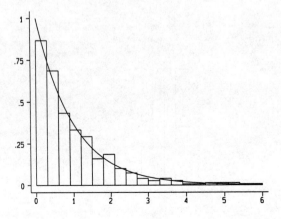

FIGURE 4.8
Members of the exponential family and a simulated data set.

of Figure 4.8. The histogram is for a sample of size 500 from the exponential curve having mean 1.

4.4.8 The Gamma Family

In the Distribution Family item, click on Gamma, and in the Option item, click on Curves. Then click on Run. You will see the graph shown at the top of Figure 4.9. The four curves here are for values of the parameter α from 1 to 4, respectively. Like the family of exponential curves, the gamma family is used to describe populations that must have positive values. In fact, one way that gamma curves arise is if we take a population having the exponential distribution with mean 1 and then obtain millions of samples of size α and calculate the sum of the elements in each sample. The histogram of the resulting sums would look like the gamma curve with parameter α. This also shows that the exponential curve for mean 1 is simply a

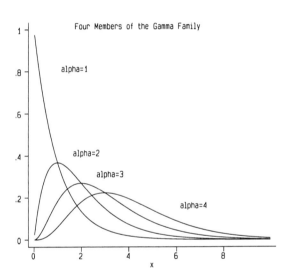
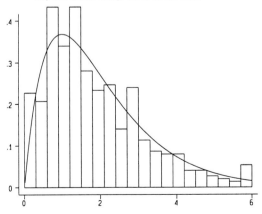

FIGURE 4.9
Members of the gamma family and a simulated data set.

special case of the gamma curve with $\alpha = 1$, which you can confirm by looking at the two figures. Finally, because of this sum-of-exponentials characteristic of gamma curves, the gamma curve for parameter α is often used to describe the amount of time it takes a machine to fail for the αth time.

From the formulas in the table at the end of this chapter, we can confirm that the mean and the variance of a population having the gamma distribution with parameter α are both equal to α. Thus, as α increases, the curve moves to the right and becomes more spread out.

Now click on Simulated-Data in the Option item and then click on Run. You will see a histogram and distribution curve similar to the one shown at the bottom of Figure 4.9. The histogram is for a sample of size 500 from the gamma curve having $\alpha = 2$.

4.4.9 The Laplace Family

In the Distribution Family item, click on Laplace, and in the Option item, click on Curves. Then click on Run. You will see the graph shown at the top of Figure 4.10. The five curves here are similar to the five curves in the normal family. The two parameters of the curves are the mean and twice the variance of the population having the curve. Although these curves are symmetrical about the mean like the normal curves, the Laplace curves are sharply "peaked" in the middle and fall away from the peak very rapidly. In fact, if you look only at the right half of each curve, you will see that it looks just like an exponential curve. Thus, the Laplace family is used in a way similar to the exponential family except for cases in which negative values are possible.

Now click on Simulated-Data in the Option item and then click on Run. You will see a histogram and distribution curve similar to the one shown at the bottom of Figure 4.10. The histogram is for a sample of size 500 from the Laplace curve having mean 0 and variance 0.5 (remember that the second parameter is twice the variance).

4.4.10 The Logistic Family

In the Distribution Family item, click on Logistic, and in the Option item, click on Curves. Then click on Run. You will see the graph shown at the top of Figure 4.11. Once again, as with the Laplace curves, we have chosen five logistic curves that are similar to the five normal curves. The first parameter a represents the mean of a population, while the second parameter b is a little more than half of the standard deviation of the population. Thus, as b increases, the variance of the population increases and the curve gets taller and narrower.

Although the logistic curves look rather similar to the normal curves, a logistic curve and a normal curve for populations having the same means and variances will not look exactly the same. The logistic family is used in many medical applications (for example, to model the growth of people).

Now click on Simulated-Data in the Option item and then click on Run. You will see a histogram and distribution curve similar to the one shown at the bottom of Figure 4.11. The histogram is for a sample of size 500 from a population having

FIGURE 4.10
Members of the Laplace family and a simulated data set.

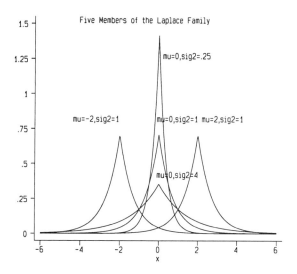

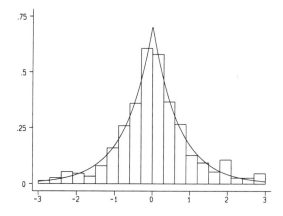

the logistic curve with $a = 0$ and $b = 1$, which means the population has mean 0 and variance approximately 3.29.

4.4.11 The Lognormal Family

In the Distribution Family item, click on Lognormal, and in the Option item, click on Curves. Then click on Run. You will see the graph shown at the top of Figure 4.12. As the graph shows, lognormal curves come in a variety of shapes. Like several other families, they are used for populations having only positive values. The parameters a and b are not the mean and variance, which, as you can see from the formulas in the table at the end of this chapter, are rather complicated functions of a and b. Notice that the curves are skewed to the right; that is, they have longer

FIGURE 4.11
Members of the logistic family and a simulated data set.

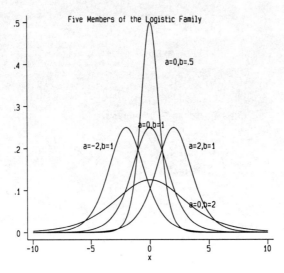

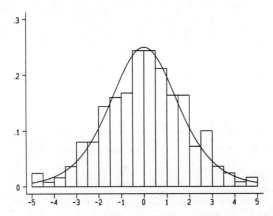

tails on the right than on the left. This is one reason lognormal curves are often used for phenomena such as rainfall intensity or the power of received radio signals.

Now click on Simulated-Data in the Option item and then click on Run. You will see a histogram and distribution curve similar to the one shown at the bottom of Figure 4.12. The histogram is for a sample of 500 elements from a population having the lognormal curve with $a = 0$ and $b = 1$. Thus, using the formulas at the end of the chapter, the mean and variance of the population are approximately 1.65 and 4.67, respectively.

4.4.12 The Pareto Family

In the Distribution Family item, click on Pareto, and in the Option item, click on Curves. Then click on Run. You will see the graph shown at the top of Figure 4.13. The four curves in the plot are the Pareto curves for parameter values a

FIGURE **4.12**
Members of the lognormal family and a simulated data set.

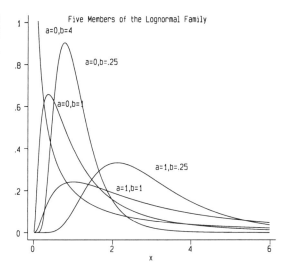

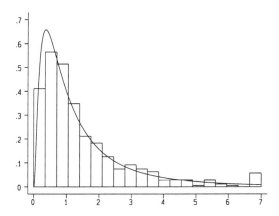

of 1, 2, 3, and 4, respectively. From the table at the end of the chapter, we see that for $a > 1$, the mean of a population having the Pareto distribution is $a/(a-1)$, which for $a = 2, 3$, and 4 are 2, 1.5, and 4/3, respectively. The Pareto curves look somewhat like the exponential curves. One unusual feature of the Pareto curves is that the values on the horizontal axis are all greater than 1. The Pareto family of curves is often used to describe quantities such as income of a population.

Now click on Simulated-Data in the Option item and then click on Run. You will see a histogram and distribution curve similar to the one shown at the bottom of Figure 4.13. The histogram is for a sample of 500 elements from a population having the Pareto distribution with $a = 2$. Using the formulas in the table at the end of the chapter, we can calculate that the population mean is 2 while the variance is essentially infinite. Note from the histogram that the values in the sample are, in fact, very spread out (look at the box at the right end of the histogram).

FIGURE 4.13
Members of the Pareto family and a simulated data set.

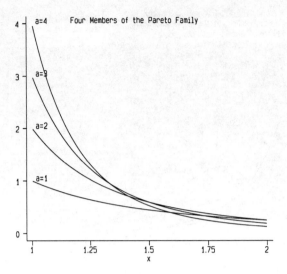

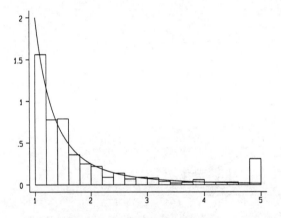

4.4.13 The Uniform Family

In the Distribution Family item, click on Uniform, and in the Option item, click on Curves. Then click on Run. You will see the graph shown at the top of Figure 4.14. The four curves in this plot all are representing the idea that all the values in a population are equally likely to occur. Each one is for a population whose values can be in different intervals. For example, the uniform(−.25, .25) curve would be for a population whose values could be anything between −0.25 and 0.25, with each value equally likely to occur.

Because the area under any distribution curve must be 1, we see that the height of one of the uniform curves is simply the reciprocal of the length of the interval. The mean of a population having the uniform curve on the interval from a to b is simply the midpoint of the interval, while the variance is the square of the length of the interval divided by 12.

FIGURE 4.14
Members of the uniform family and a simulated data set.

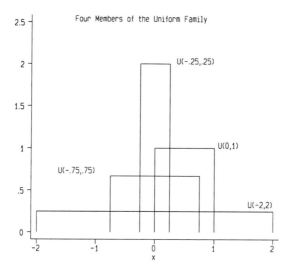

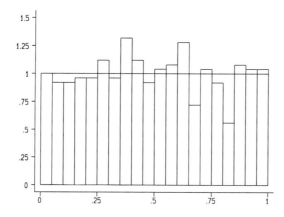

Now click on Simulated-Data in the Option item and then click on Run. You will see a histogram and distribution curve similar to the one shown at the bottom of Figure 4.14. The histogram is for a sample of 500 elements from a population having a uniform distribution on the interval from 0 to 1. Thus, the mean is 0.5 and the variance is 1/12.

People are often surprised at how "ununiform" a histogram looks even when the population being sampled from is known to be uniformly distributed. Be sure to click repeatedly on Run so you can observe this phenomenon.

4.4.14 The Weibull Family

In the Distribution Family item, click on Weibull, and in the Option item, click on Curves. Then click on Run. You will see the graph shown at the top of Figure 4.15. The four curves in this plot are for values of parameter c of 1, 2, 3, and 4, respectively. The mean and variance are, in fact, complicated functions of c. The Weibull family of curves is a generalization of the family of exponential curves; it is also used to describe lifetimes and tensile strength of materials, as well as in statistical quality control.

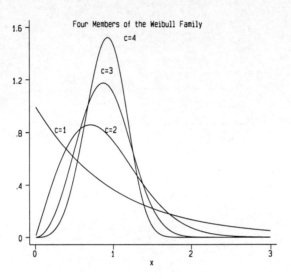
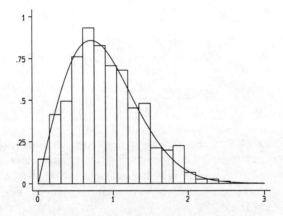

FIGURE **4.15** Members of the Weibull family and a simulated data set.

Now click on Simulated-Data in the Option item and then click on Run. You will see a histogram and distribution curve similar to the one shown at the bottom of Figure 4.15. The histogram is for a sample of 500 members of a population having the Weibull distribution with $c = 2$. Using the formulas in the table at the end of the chapter, we see that the mean and variance of the population are approximately 0.8862 and 0.2146, respectively.

4.5 Summary

In our tour of this lab, we learned several things about population distributions:

1. At least 14 different families of distributions can be used to describe continuous populations.
2. There is such a wealth of diversity in these curves that the pattern of occurrence of almost any scientific phenomenon can be well described by some curve in some family.
3. One main goal of statistical inference from data is to find the best possible curve from this wealth of curves to describe the population.

4.6 Lab Exercises

1. What happens to the height and width of a normal curve as the variance (the second parameter) decreases?
2. Which has heavier tails, t curves or the standard normal curve?
3. Print out the chi-square graph and write the degrees of freedom above each curve. Does the chi-square curve get wider or narrower as the degrees of freedom increase?
4. Does the F curve get more or less symmetrical as degrees of freedom increase?
5. What would you guess the population median to be if a population had the beta distribution with $p = 2$ and $q = 9$? How about $p = 3$ and $q = 3$?
6. How does the mean of populations having the beta curve with $p = 2$ and $q = 9$ compare with that having $p = 9$ and $q = 2$. What about the population variances?
7. Which families have curves that can be used to describe populations that can have negative values? How about populations that must have all positive values?
8. For what values on the horizontal axis do the standard normal and standard Cauchy curves cross?
9. How do the gamma curve with $\alpha = 2$ and the lognormal curve with $a = 0$ and $b = 1$ differ?
10. For which families are all the curves we have plotted symmetrical?
11. What are four families of distributions that have at least one member whose shape is similar to the members of the exponential family?
12. What is the basic difference in shape of the members of the Laplace family versus the logistic family?
13. Which of the four members of the Pareto family we graphed should be used for an economy that has a very unequal distribution of income?
14. Which of the four uniform curves that we plotted has the smallest variance?
15. Print four uniform histograms. How much do they seem to vary?

4.7 Formulas for Continuous Distributions, Means, and Variances

In this section, we list the formulas for the curves for all 14 families of distributions, as well as their mean and variance, in alphabetical order. We use the symbols $f(x)$, μ_X, and σ_X^2 to represent the three quantities for each family.

Beta with parameters p and q

$$f(x) = \frac{\Gamma(p+q)}{\Gamma(p)\Gamma(q)} x^{p-1}(1-x)^{q-1}, \qquad 0 \le x \le 1, \quad p > 0, q > 0$$

$$\mu_X = \frac{p}{p+q}, \qquad \sigma_X^2 = \frac{pq}{(p+q)^2(p+q+1)}$$

Cauchy with parameters a and b

$$f(x) = \frac{1}{\pi b} \frac{1}{1 + ((x-a)/b)^2}, \qquad -\infty < x < \infty, \quad -\infty < a < \infty, b > 0$$

μ_X and σ_X^2 do not exist.

Chi-square with ν degrees of freedom

$$f(x) = \frac{1}{\Gamma(\nu/2) 2^{\nu/2}} x^{(\nu/2)-1} e^{-x/2}, \qquad x > 0, \quad \nu = 1, 2, 3, \ldots$$

$$\mu_X = \nu, \qquad \sigma_X^2 = 2\nu$$

Exponential with parameter μ

$$f(x) = \frac{1}{\mu} e^{-x/\mu}, \qquad x > 0, \quad \mu > 0$$

$$\mu_X = \mu, \qquad \sigma_X^2 = \mu^2$$

F with ν_1 and ν_2 degrees of freedom

$$f(x) = \frac{\Gamma((\nu_1+\nu_2)/2)}{\Gamma(\nu_1/2)\Gamma(\nu_2/2)} \left(\frac{\nu_1}{\nu_2}\right)^{\nu_1/2} \frac{x^{(\nu_1-2)/2}}{\left(1 + \frac{\nu_1}{\nu_2}x\right)^{(\nu_1+\nu_2)/2}}, \qquad x > 0, \quad \nu_1, \nu_2 = 1, 2, 3, \ldots$$

$$\mu_X = \frac{\nu_2}{\nu_2 - 2}, \quad \nu_2 > 2, \qquad \sigma_X^2 = 2\left(\frac{\nu_2}{\nu_2-2}\right)^2 \frac{(\nu_1+\nu_2-2)}{\nu_1(\nu_2-4)}, \quad \nu_2 > 4$$

Gamma with parameter α

$$f(x) = \frac{1}{\Gamma(\alpha)} x^{\alpha-1} e^{-x}, \qquad x > 0, \quad \alpha > 0$$

$$\mu_X = \alpha, \qquad \sigma_X^2 = \alpha$$

Laplace with parameters μ and σ^2

$$f(x) = \frac{1}{2\sigma} e^{-|x-\mu|/\sigma}, \qquad -\infty < x < \infty, \quad -\infty < \mu < \infty, \quad \sigma^2 > 0$$

$$\mu_X = \mu, \qquad \sigma_X^2 = 2\sigma^2$$

Logistic with parameters a and b

$$f(x) = \frac{1}{b} \frac{e^{-(x-a)/b}}{\left(1 + e^{-(x-a)/b}\right)^2}, \qquad -\infty < x < \infty, \quad -\infty < a < \infty, \quad b > 0$$

$$\mu_X = a, \qquad \sigma_X^2 = \frac{\pi^2 b^2}{3}$$

Lognormal with parameters a and b

$$f(x) = \frac{1}{\sqrt{2\pi b}} \frac{e^{-(\log x - a)^2 / 2b}}{x}, \qquad x > 0, \quad -\infty < a < \infty, \quad b > 0$$

$$\mu_X = e^{a+b/2}, \qquad \sigma_X^2 = e^{2(a+b)} - e^{2a+b}$$

Normal with parameters μ and σ^2

$$f(x) = \frac{1}{\sqrt{2\pi}\sigma} e^{-(x-\mu)^2 / 2\sigma^2}, \qquad -\infty < x < \infty, \quad -\infty < \mu < \infty, \quad \sigma^2 > 0$$

$$\mu_X = \mu, \qquad \sigma_X^2 = \sigma^2$$

Pareto with parameter a

$$f(x) = \frac{a}{x^{a+1}}, \qquad x > 1, \quad a > 0$$

$$\mu_X = \frac{a}{a-1}, \quad a > 1, \qquad \sigma_X^2 = \frac{a}{(a-1)^2 (a-2)}, \quad a > 2$$

t with ν degrees of freedom

$$f(x) = \frac{\Gamma((\nu+1)/2)}{\Gamma(\nu/2)} \frac{1}{\sqrt{\nu\pi}} \frac{1}{(1 + x^2/\nu)^{(\nu+1)/2}}, \qquad -\infty < x < \infty, \quad \nu = 1, 2, 3, \ldots$$

$$\mu_X = 0, \quad \nu > 1, \qquad \sigma_X^2 = \frac{\nu}{\nu - 2}, \quad \nu > 2$$

Uniform with parameters a and b

$$f(x) = \frac{1}{b-a}, \quad a < x < b$$

$$\mu_X = \frac{a+b}{2}, \quad \sigma_X^2 = \frac{(b-a)^2}{12}$$

Weibull with parameter c

$$f(x) = cx^{c-1}e^{-x^c}, \quad x \geq 0, \quad c > 0$$

$$\mu_X = \Gamma(1+1/c), \quad \sigma_X^2 = \Gamma(1+2/c) - \Gamma^2(1+1/c)$$

5

Sampling From 0–1 Populations

Chapter 4 considered continuous populations, that is, populations that can contain many possible values. In this chapter, we consider sampling from populations whose elements can take on only two possible values.

5.1 Introduction

In many circumstances, such as political polls or feasibility studies, the parameter of interest is the proportion of favorable responses in the population. To study this, a random sample is chosen, and the favorable responses, or "successes," are counted. Numerically, we record a success as a 1 and a failure as a 0. This lab will guide you through an examination of many different aspects of sampling from a "0–1 population."

5.1.1 Some Basic Ideas

Suppose we have a 0–1 population having N elements and the proportion of 1's is π. A random sample of size n is chosen, and the number of 1's is counted. There are two ways that such a sample can be obtained:

1 *Sampling with replacement:* In this method, we randomly select an element from the population, record whether it is a 0 or a 1, and return it to the population. Then we repeat this process until we have selected n of the elements.

2 *Sampling without replacement:* In this method, we do not replace each element before selecting the next.

The Binomial and Hypergeometric Distributions

If we take a sample of size n with replacement and count the number X of 1's in the sample, then the distribution of X is called the *binomial distribution* with parameters n and π. If we want to calculate the probability that the number of 1's in our sample is equal to x, the *probability distribution function* is

$$P(X = x) = \binom{n}{x} \pi^x (1 - \pi)^{n-x}, \quad x = 0, 1, \ldots, n.$$

If sampling is done with replacement, we can easily calculate the mean number (or *expected number*) of 1's using

$$\mu_X = E(X) = n\pi.$$

(Notice that if the expected number of 1's is $n\pi$, then the expected number of 0's would be $n(1 - \pi)$.) The variance of the number of 1's is

$$\sigma_X^2 = \text{Var}(X) = n\pi(1 - \pi).$$

Sometimes, calculating probabilities from a binomial distribution can be time-consuming or tedious, or both, and tables are not available (for example, when $n = 327$ and $\pi = 0.0325$ and we want $P(X < 259)$). When this happens, binomial probabilities can be approximated using other probability distribution functions. If n is large and π is small, the probability distribution function called the Poisson distribution can be used. If the smaller of $n\pi$ (the expected number of 1's) and $n(1 - \pi)$ (the expected number of 0's) is at least 5, then the normal distribution may be used.

Now suppose we are taking a sample of size n without replacement and are still interested in the number X of 1's in the sample. Denote the number of 1's in the population by M. Now the distribution of X is called a *hypergeometric distribution* with parameters $N, n,$ and π. (Notice that $\pi = M/N$.) The hypergeometric probability distribution function is

$$P(X = x) = \frac{\binom{M}{x}\binom{N-M}{n-x}}{\binom{N}{n}}, \quad x = 0, 1, \ldots, n.$$

The mean (or expected) number of 1's from a sample drawn without replacement is

$$\mu_X = E(X) = n\pi.$$

The variance is

$$\sigma_X^2 = \text{Var}(X) = \left(\frac{N-n}{N-1}\right) n\pi(1 - \pi).$$

Notice that the mean of the number of 1's when sampling with replacement is the same as when the sampling is done without replacement. The variances differ only by a factor of $C = (N - n)/(N - 1)$, called the *finite population correction factor*. Later on, we will see that the closer the finite population correction factor is to 1, the more alike are the binomial and hypergeometric distributions.

The Negative Binomial Distribution

A third circumstance that arises when sampling with replacement from a 0–1 population is when we are interested in the number X of elements drawn from the population until the nth 1 is observed. In this case, the distribution of X is called the *negative binomial distribution* with parameters n and π. For a negative binomial distribution, we calculate the probability that it requires x draws to get the nth 1. The negative binomial probability distribution function is

$$P(X = x) = \binom{x - 1}{n - 1} \pi^n (1 - \pi)^{x-n}, \quad x = n, n + 1, n + 2, \ldots.$$

We can use

$$\mu_X = E(X) = \frac{n}{\pi}$$

to calculate the mean (or expected) number of draws needed to observe the nth 1. To calculate the variance of the number of necessary draws, we use

$$\sigma_X^2 = \text{Var}(X) = \frac{n(1 - \pi)}{\pi^2}.$$

5.2 Objectives

1. Illustrate the distribution of the number of 1's from a random sample of size n chosen with replacement (a distribution known as the binomial distribution with parameters n and π).

2. Illustrate the distribution of the number of 1's from a random sample of size n chosen without replacement (a distribution known as the hypergeometric distribution with parameters N, n, and π).

3. Visualize the effects of N, n, and π on the two distributions.

4. Illustrate the distribution of the number of individuals drawn from the population until the nth 1 is observed (a distribution known as the negative binomial distribution with parameters n and π).

5. Visualize the effects of n and π on the shape of the distribution.

6. Use the Poisson distribution to approximate the binomial distribution and visualize the effect of the values of n and π on this approximation.

7. Use the normal distribution to approximate the binomial distribution and visualize the effect of the values of n and π on this approximation.

5.3 Description

To begin the lab, select Sampling from 0-1 Populations from the Labs menu. The following submenu appears with a choice of three concept labs:

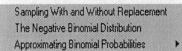

5.3.1 Sampling With and Without Replacement

Choosing this lab opens the dialog box shown in Figure 5.1. Toward the bottom left-hand corner of that window are three boxes where you specify desired values for the population size N, the population proportion of 1's pi, and the sample size n. Initial values are N = 50, pi = 0.50, and n = 10. Other values for these parameters may be specified either by clicking on the desired values shown or by typing them in the box. The values for N and pi are not restricted to those shown.

To the right of the three boxes are three buttons:

1. Run: Clicking on this button opens a graphics window that shows a graph of the two distributions (see Figure 5.2). The narrow bars drawn in red represent the hypergeometric distribution with values of N, pi, and n as specified in the dialog box. The wider bars, drawn in blue, represent the binomial distribution with values of n and pi as specified in the dialog box.

2. Close: Clicking on this button closes the lab and returns you to the StatConcepts menus.

3. Help: Clicking on this button opens a help window containing information about the lab.

FIGURE 5.1
Dialog box of the *Sampling With and Without Replacement* Lab.

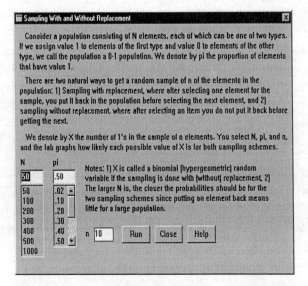

FIGURE 5.2
Example of the *Sampling With and Without Replacement* Lab.

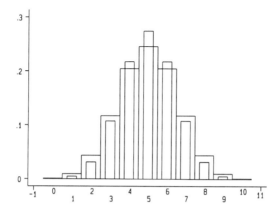

5.3.2 The Negative Binomial Distribution

Choosing this concept lab will open the dialog box shown in Figure 5.3. Toward the bottom left-hand corner of that window are two boxes where you can specify desired values for the sample size n and the population proportion of 1's pi. Initial values are n = 5 and pi = 0.50. Other values for these parameters may be specified either by clicking on the desired values shown or by typing them in the box. The values for n and pi are not restricted to those shown.

To the right of the two boxes are three buttons:

1 Run: Clicking on this button opens a graphics window that shows a graph of the negative binomial distribution for the values specified for n and pi (see Figure 5.4).

FIGURE 5.3
Dialog box for the *Negative Binomial* Lab.

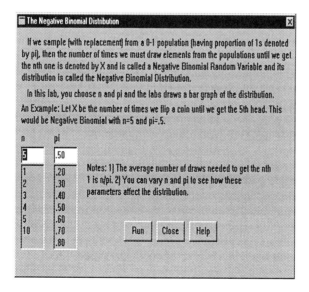

FIGURE 5.4
Example of the *Negative Binomial* Lab.

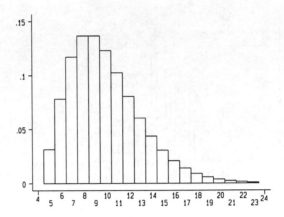

2 Close: Clicking on this button closes the lab and returns you to the StatConcepts menus.

3 Help: Clicking on this button opens a help window containing information about the lab.

5.3.3 Approximating Binomial Probabilities

Sometimes, calculating binomial probabilities can be time-consuming or tedious, or both, and tables are not available. In such circumstances, it is possible to use other distributions to approximate binomial probabilities. If you choose Approximating Binomial Probabilities, the following submenu appears:

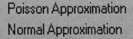

Poisson Approximation to Binomial

Choosing this concept lab will open the dialog box shown in Figure 5.5, which contains a brief discussion on how the Poisson distribution arises and how it may be used to approximate the binomial distribution. Toward the bottom left-hand corner of that window are two boxes where you can specify desired values for n and pi. Initial values are n = 50 and pi = 0.10. Other values for these parameters may be specified either by clicking on the desired values shown or by typing them in the boxes. The values of n and pi are not restricted to those listed.

To the right of the boxes are three buttons:

1 Run: Clicking on this button opens a graphics window that shows a graph of the two distributions (see Figure 5.6). The wider bars drawn in blue represent the binomial distribution for the specified values of n and pi. The narrower bars drawn in red represent the value of the Poisson approximation.

FIGURE **5.5**
Dialog box for the *Poisson Approximation to Binomial Probabilities* Lab.

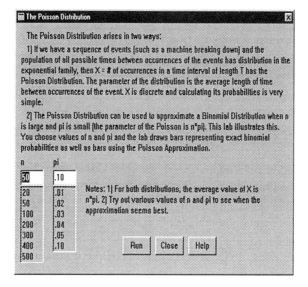

FIGURE **5.6**
Example of the *Poisson Approximation to Binomial* Lab.

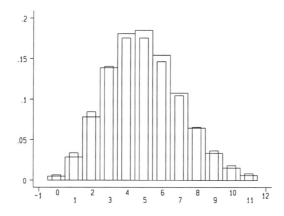

2 `Close`: Clicking on this button closes the lab and returns you to the StatConcepts menus.

3 `Help`: Clicking on this button opens a help window containing information about the lab.

Normal Approximation to Binomial

Choosing this concept lab will open the dialog box shown in Figure 5.7, which contains a brief discussion on when it is appropriate to use the normal distribution to approximate binomial probabilities. Toward the bottom left-hand corner are two boxes where you can specify desired values for n and pi. For this lab, the values of n must be between 2 and 49. Initial values are n = 10 and pi = 0.50.

FIGURE **5.7**
Dialog box for the *Normal Approximation to Binomial Probabilities* Lab.

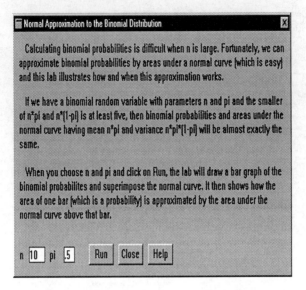

To the right of the boxes are three buttons:

1 `Run`: When you click on `Run`, a graphics window opens that shows a graph of the binomial probability distribution for the specified values of `n` and `pi` (see Figure 5.8). The normal curve approximation is superimposed. It also shows how the area of one bar from the binomial distribution is approximated by the area under the normal curve above that bar.

2 `Close`: Clicking on this button closes the lab and returns you to the StatConcepts menus.

3 `Help`: Clicking on this button opens a help window containing information about the lab.

FIGURE **5.8**
Example of the *Normal Approximation to Binomial* Lab.

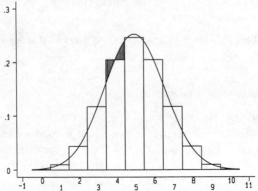

5.4 Guided Tour of the Lab

There are many stops to make and many sights to see in this tour. Each stop has at least a few notable attractions. Our tour will be a long but enjoyable one as we see all there is to see about sampling from 0–1 populations.

5.4.1 Sampling With and Without Replacement

Let's begin the tour by studying the distribution of the number of 1's in a sample of size n when the sample is chosen from a population in which the proportion of 1's is π. Recall that we are interested in the number X of 1's in our sample. When we sample with replacement, X has a binomial distribution with parameters n and π; when we sample without replacement, X has a hypergeometric distribution with parameters N, n, and π.

Begin the *Sampling With and Without Replacement* Lab under `Sampling from 0-1 Populations` from the Labs menu. There are a few things you should notice. First, for the initial values N = 50, n = 10, and pi = 0.50, both distributions are symmetrical and are centered about their common mean $n\pi = (10)(0.50) = 5$. We will see shortly that as the values of n and π change, so will the shapes of these distributions. Second, for small (0, 1, 2, 3) and large (7, 8, 9, 10) values of x, the probabilities (heights of the bars) for the hypergeometric distribution are less than the probabilities for the binomial distribution, but at the center of the distributions ($x = 4, 5, 6$), the binomial probabilities are less than the hypergeometric distribution. As we continue, we will observe that the differences in these two distributions diminish as the size of the population N increases.

The Effect of Changing n and π

First, let's see what happens to the shapes of the distributions as the values of n and pi change. Changing n from 10 to 5 will result in the graph shown in Figure 5.9. Compare the distributions in this graph with those in Figure 5.2. What do you notice about the differences in these distributions? You should see that there are fewer bars in the new distributions. When n = 10, there were 11 bars; now that n = 5, there are 6. This is because there is exactly one bar for every possible value for the number of 1's in a sample of size n. Therefore, when n = 10, we can observe zero 1's, one 1, two 1's, and so on all the way up to ten 1's. But when n = 5, we can observe only up to five 1's.

Also notice that the center of the distribution is still the mean $n\pi = (5)(0.50) = 2.5$; that is, we expect to observe two and a half 1's in our sample of size 5. At first glance, this doesn't seem to be a reasonable value for the average number of 1's that we would count. After all, if we're counting the *number* of 1's, how does an average of 2.5 make any sense? But if we consider the definition of the mean of a sampling distribution, we see that this value is actually quite reasonable. A population with $\pi = 0.50$ has many, many samples of size 5. And, for every sample of size 5 that we choose, we can count the number of 1's. If we could choose all possible samples of size 5 and then average the number of 1's from each sample, the result would be 2.5.

FIGURE 5.9
Binomial and hypergeometric distributions for $N = 50$, $n = 5$, and `pi` = 0.50.

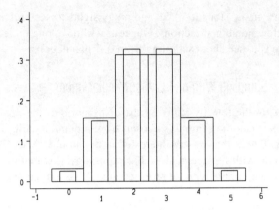

We can make this same statement for any sample size *n* and any value of π. That is, we can say that the mean number of 1's is the average of the number of 1's from every possible sample of size *n*. Finally, note that the mean of a distribution is the center of its mass. In other words, the mean is the point where the distribution would balance if it was placed on a fulcrum. Thus, all binomial and hypergeometric distributions balance at $n\pi$.

Now we're going to see what happens to the shapes of the distributions when the value of π changes and *n* stays the same. Changing the value of n back to 10 and the value of `pi` from 0.50 to 0.02 will result in a graph like the one shown in Figure 5.10. Compare this to the graph in Figure 5.2. Notice how the shapes of both of the new distributions have become severely skewed to the right. This is because the probability of observing a 1 has become small. This means that the chance of observing a small number of 1's is large and the chance of observing a large number of 1's is small. (Remember, the height of a bar represents the probability of observing

FIGURE 5.10
Binomial and hypergeometric distributions for $N = 50$, $n = 5$, and `pi` = 0.02.

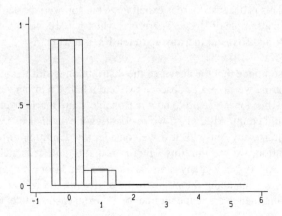

the number of 1's shown on the horizontal axis under that bar.) The mean of these distributions is $n\pi = (10)(0.02) = 0.20$. Does it appear that the distribution would balance at 0.20?

Figure 5.11 shows a graph of the binomial and hypergeometric distributions for pi $= 0.20$ and n $= 5$. You should generate this graph, too. The distributions are still skewed to the right, but less severely. The mean is $n\pi = (5)(0.20) = 1$. As the shape of the distribution shifts to the right, so does the mean. Now run the lab again for pi $= 0.50, 0.80$, and 0.98 (see Figures 5.2, 5.12, and 5.13). What happens to the shapes of the distributions as the value of pi changes? (It should be symmetrical at pi $= 0.50$ and become more and more skewed to the left for pi $= 0.80$ and 0.98.) The values of the mean for each of these three values of pi are 5, 8, and 9.8, respectively. Notice how as the distribution becomes more and more skewed to the left, the value of the mean increases.

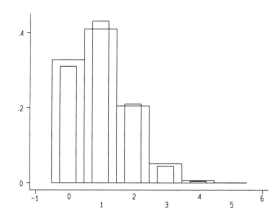

FIGURE 5.11
Binomial and hypergeometric distributions for N = 50, n = 5, and pi = 0.20.

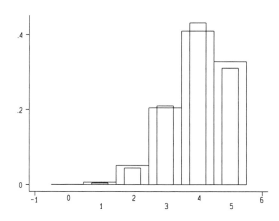

FIGURE 5.12
Binomial and hypergeometric distributions for N = 50, n = 5, and pi = 0.80.

FIGURE 5.13
Binomial and hypergeometric distributions for N = 50, n = 5, and pi = 0.98.

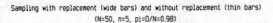
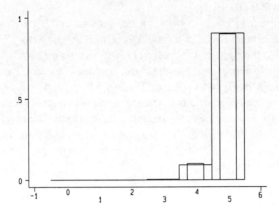

Effect of Changing N

Recall that the hypergeometric distribution is the distribution of the number of 1's when sampling without replacement from a population of size N. Therefore, by changing the value of N, we also change the shape of the hypergeometric distribution. In particular, the larger the value of N becomes relative to the value of n, the more the hypergeometric distribution begins to resemble the binomial distribution (when sampling with replacement).

To see this, generate the graph shown in Figure 5.14 by letting N = 500, with n = 10 and pi = 0.50. Compare this figure to Figure 5.2, where N = 50. Notice that the differences in the heights of the bars of the distributions for N = 50 are much larger than for N = 500. The value of the finite population correction factor when N = 50 and n = 10 is $C = (N - n)/(N - 1) = (50 - 10)/(50 - 1) = 0.8163$. When N = 500 and n = 10, this becomes $C = 0.9820$.

FIGURE 5.14
Binomial and hypergeometric distributions for N = 500, n = 10, and pi = 0.50.

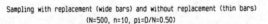
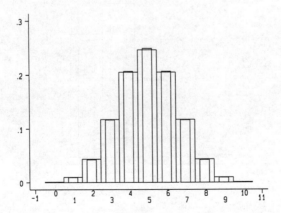

Now let's see what happens when the value of the sample size *n* is close to the value of the population size *N*. Let N = 50 again but change n to 40 (see Figure 5.15). For these values of N and n, $C = 0.2041$. Notice how different the binomial and hypergeometric distributions are! Finally, leave n = 40 but change N to 500 (see Figure 5.16). Now $C = 0.9218$. Again, there is very little difference in the heights of the bars for the two distributions.

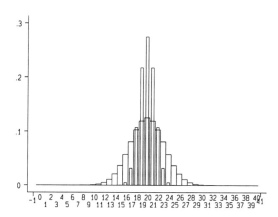

FIGURE **5.15**
Binomial and hypergeometric distributions for N = 50, n = 40, and pi = 0.50.

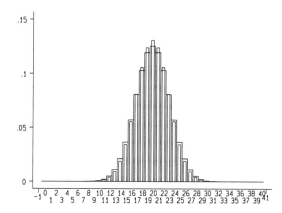

FIGURE **5.16**
Binomial and hypergeometric distributions for N = 500, n = 40, and pi = 0.50.

One final note: You should have noticed that changing the value of *N* had no effect on the shape of the binomial distribution. Because binomial probabilities are based on a sample selected with replacement, the probability that an individual will be selected does not change from one selection to the next. Thus, the probability of observing *x* 1's from a sample of size *n* is unaffected by the population size. The distribution is a graph of these probabilities.

5.4.2 The Negative Binomial Distribution

Suppose we are sampling with replacement, but instead of being concerned with the number of 1's in a sample of size n, we are interested in the number of draws needed until the nth 1 is observed. In this case, the random variable X is the number of draws necessary, and the distribution of X is called the *negative binomial distribution with parameters n and π* (where π is the proportion of 1's in the population). Our tour will now guide us through the behavior of this new distribution.

A graph of the negative binomial distribution for n = 5 and pi = 0.50 is shown in Figure 5.4. There are a few things you should be aware of in this graph:

1. The scale on the horizontal axis begins at the value n = 5. This is because the horizontal axis represents the number of draws necessary to obtain the fifth 1, and we cannot get five 1's in fewer than five draws. A graph of the negative binomial distribution will always begin at n.

2. Although 19 bars are shown, there is potentially an infinite number of bars, because theoretically we could draw an infinite number of elements from our population and never observe five 1's. However, the longer we continue to draw elements from the sample, the more unlikely it becomes that we would not have observed the fifth 1. Because the height of the bars represents the probability of observing the fifth 1 on the nth draw, at some point the bars become smaller and smaller (in this graph, they start getting smaller at $n = 10$). We can visually detect the height of these bars only up to $n = 23$. After that, the chance that we're still trying to get the fifth 1 is so small that even though the height is not zero, it is too small to be seen on the graph.

3. The mean of a negative binomial distribution is given by $\mu_X = n/\pi = 5/0.50 = 10$. The mean is interpreted as follows. We draw a sample with replacement until we get the fifth 1 and record the number of elements drawn. Then we repeat this procedure many, many times. If we averaged the number of elements drawn from each sample we chose (from each time we repeated the procedure), that average would be 10. Thus, in the long run, we expect to draw 10 elements with replacement until the fifth 1 is observed. Physically, the negative binomial distribution will balance at 10.

Effect of Changing n and π

Leaving pi = 0.50 and decreasing n from 5 to 3 will result in a graph like the one shown in Figure 5.17. Compare this figure to Figure 5.4 and note these differences:

1. The horizontal axis now begins at 3 because it is possible to observe the n = third 1 by the third draw. When n = 5, we need at least five draws to observe the fifth 1.

2. There appear to be fewer bars for n = 3 than for n = 5. This is because now we only want to get to the third 1. This should happen in a fewer number of draws, and, in fact, we see that behavior reflected in the lesser number of bars in the graph of the distribution. But don't forget that in both cases, the number of bars

FIGURE **5.17**
Negative binomial distribution with n = 3 and pi = 0.50.

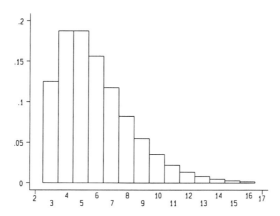

continues on and on. It's just that the heights get smaller faster for smaller values of n.

3 The mean for n = 3 is $\mu_X = n/\pi = 3/0.50 = 6$; in the long run, we expect to draw only six times to get the third 1. And visually, we can see that the distribution would balance at 6.

Now, changing n back to 5 and decreasing pi to 0.20 will result in a graph like the one shown in Figure 5.18. Again, we will be comparing this figure to Figure 5.4. Probably, the first thing you notice is the large increase in the number of visible bars. The reason for this is easily understood if you reflect for a moment on the meaning of π, which we have decreased from 0.50 to 0.20. The value of π tells us the proportion of 1's present in the population. If this value is smaller, it stands to reason that it will take a longer time (or a larger number of draws) to finally get to the fifth 1. Therefore, we see more bars. This behavior is also reflected in the value of the mean for this

FIGURE **5.18**
Negative binomial distribution with n = 5 and pi = 0.20.

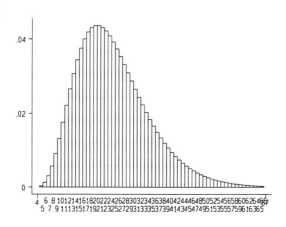

new negative binomial distribution. Here $\mu_X = n/\pi = 5/0.20 = 25$. Now that `pi` = 0.20, we expect to need 25 draws to obtain the fifth 1 (as opposed to the much smaller 10 draws when `pi = 0.50`).

5.4.3 Approximating Binomial Probabilities

Again, calculating probabilities from a binomial distribution sometimes can be time-consuming or tedious, or both, and tables are not available to make the job easier. When this happens, if certain conditions are met, we can use other distribution functions to approximate binomial probabilities. The remainder of our tour will assess how well two popular methods for approximating binomial probabilities perform.

The Poisson Approximation

Before we discuss how the Poisson distribution function is used to approximate binomial probabilities, a word about the Poisson distribution itself is in order. The Poisson distribution arises in two ways:

1. Suppose we have an event that occurs over and over again (for example, buses stopping at a bus stop or a machine breaking down), and the population of all possible times between occurrences of those events has a distribution from the family of exponential distributions (see the *How Are Populations Distributed?* Lab). Then the number X of occurrences of the event in a time interval of length T has a Poisson distribution. The parameter for the Poisson distribution is the product of the length T of the time interval and the average length of time between occurrences of the event. The distribution of X is discrete, and calculating the probability that X is equal to some particular value x is very simple.

2. If n is large and π is small, then the Poisson distribution has the parameter $n\pi$ (the same as the mean of the binomial distribution) and gives a very good approximation of the binomial distribution. The lab we are about to tour will show you just how good this approximation can be.

Run the lab by selecting `Poisson Appproximation` from the *Approximating Binomial Probabilities* menu in the *Sampling From 0–1 Populations* Lab. The initial values for n and `pi` are 50 and 0.10, respectively (see Figure 5.6). Notice that even at these somewhat moderate values, the differences between the heights of the blue bars (the binomial distribution) and the red bars (the Poisson distribution) are very small. As a matter of fact, the largest difference occurs at the mean, and it is only 0.0094! Now let's see what happens to this approximation as the values of n and π change.

First, change n to 20 and run the lab. The resulting graph will look like the one shown in Figure 5.19. Compare this graph to the one in Figure 5.6. Notice again that the approximation is worse at $\mu_X = n\pi = (20)(0.10) = 2$ than at any other place. Still, the approximation is quite good. The difference in the heights of the bars is only 0.0145. (At $X = 0$, the difference is 0.0137.) Thus, even for this "small" value of n and this moderate value of `pi`, the approximation is still pretty good. Now change n to 200 to get the graph shown in Figure 5.20. Only the bars with heights large enough to be visible are shown. Notice that using the Poisson distribution to approximate the

FIGURE **5.19**
Poisson approximation to the binomial distribution for n = 20 and pi = 0.10.

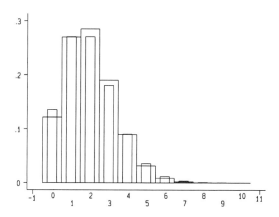

FIGURE **5.20**
Poisson approximation to the binomial distribution for n = 200 and pi = 0.10.

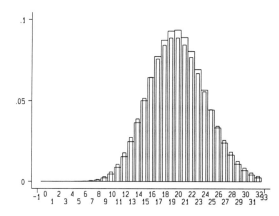

binomial distribution is extremely accurate. In fact, the largest difference (occurring at the mean $n\pi = (200)(0.10) = 20$) is only 0.0043. Therefore, we have seen that as the sample size n increases, even for moderate values of π, the Poisson approximation becomes a better and better tool for calculating binomial probabilities.

Now let's examine the approximation for changing values of π by changing n back to 50 and pi to 0.05, which will result in the graph shown in Figure 5.21. Compare this figure to Figure 5.6. Notice how much closer the Poisson distribution is to the binomial distribution. The largest difference is only 0.0061 (which occurs for $x = 3$). When you compare this to the largest difference (which was 0.0094 when n = 50 and pi = 0.10), it is easy to see that the smaller the value of π, the better the Poisson distribution becomes at approximating the binomial distribution.

One final note: Throughout this discussion, you may have noticed that we reported exact differences in the heights of the bars. These numbers were obtained by choosing Statistical Tables under the Calculator function in the main StataQuest menu. The Statistical Tables pull-down menu, shown below, has options

74 Chapter 5 Sampling From 0–1 Populations

FIGURE **5.21**
Poisson approximation to the binomial distribution for n = 50 and pi = 0.05.

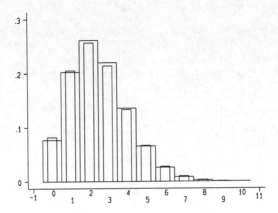

that will allow us to calculate probabilities from both the binomial and Poisson distributions.

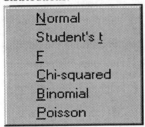

For example, to calculate the differences in the heights of the bars at 5 when n = 50 and pi = 0.10, you would follow these steps:

1. Choose Binomial from the Statistical Tables pull-down menu.
2. Complete the dialog box as shown below and click on Run. This calculates $P(X = 5)$ from the binomial distribution with $n = 50$ and $\pi = 0.10$.

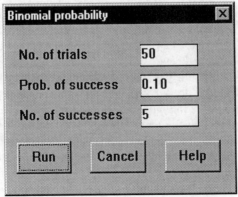

3. Now click on Cancel in the Binomial probability dialog box and choose Poisson from the Statistical Tables pull-down menu.

Complete the dialog box as shown below and click on Run. (The value of lambda is simply equal to $n\pi$.) This calculates $P(X = 5)$ from the Poisson distribution with parameter $n\pi = 5$.

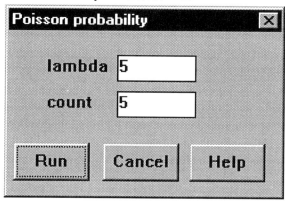

4 Click on Cancel in the Poisson probability dialog box. The *Stata Results* window will look something like what is shown in Figure 5.22. To get the difference in the heights of the bars, simply subtract the probabilities 0.1849 (for the binomial distribution) and 0.1755 (for the Poisson distribution). This gives us the difference in the heights of the bars; that is, $0.1849 - 0.1755 = 0.0094$.

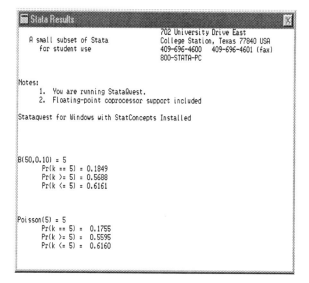

FIGURE **5.22**
Stata Results window for calculating binomial and Poisson probabilities.

The Normal Approximation

The final part of the tour will explore another method of approximating binomial probabilities. Here, we will see how the normal distribution (a continuous distribution) may be used to approximate the binomial distribution (a discrete distribution). The normal distribution may be used to approximate the binomial distribution when the smaller of $n\pi$ (the expected number of 1's) and $n(1-\pi)$ (the expected number of 0's) is at least 5.

Run the lab by choosing Normal Approximation from the pull-down menu of *Approximating Binomial Probabilities*. You will see a dialog box explaining how to use this lab and a graph like the one shown in Figure 5.8. Before we examine the behavior of the approximation as the values of n and π change, let's examine this graph. Here, we see superimposed the binomial distribution for $n = 10$ and $\pi = 0.50$ and the normal distribution with mean $\mu = n\pi = 5$ and variance $\sigma^2 = n\pi(1-\pi) = 2.5$. The bar corresponding to $X = 4$ has been chosen to illustrate how this normal distribution may be used to approximate binomial probabilities.

If we used the binomial probability distribution function to calculate $P(X = 4)$, we would get $P(X = 4) = 0.2051$. However, for the normal distribution, $P(X = 4) = 0$, because the normal distribution is continuous. Therefore, to use the normal distribution to approximate this probability, we calculate $P(3.5 < X < 4.5) = 0.2045$. This approximation works well because even though we "chop off" the area of the bar for the binomial distribution from 3.5 to 4, we "add in" almost the same amount of area under the curve from 4 to 4.5. The difference in the exact binomial probability (0.2051) and the approximated probability (0.2045) is equal to the difference in the "chopped off" area and the "added in" area. This difference is only 0.0006!

Now let's see what happens as the values of n and π change. Change the value of n from 10 to 40. Check to make sure that the smaller of $n\pi$ and $n(1-\pi)$ is at least 5 (it is, but you should check). We see in Figure 5.23 that the area of the shaded portions is getting smaller. In fact, the difference in the shaded areas (which is the same thing as the difference between the exact probability and the approximated probability) is only $0.1196 - 0.1194 = 0.0002$. Therefore, as the sample size increases, the approximation moves closer and closer to the probability we would obtain if we were using the binomial distribution itself.

Now, with n = 40, change the value of pi to 0.80. In the resulting graph, shown in Figure 5.24, we see the implications of the value of pi moving away from 0.50. Note that when pi = 0.50, the binomial distribution, like the normal distribution, is

FIGURE 5.23
Normal approximation to binomial probabilities for n = 40 and pi = 0.50.

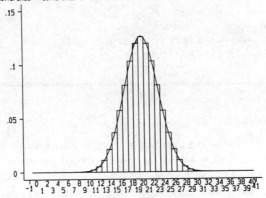

Approximate the area of a bar by the area under its part of normal curve
Ex: P(X=19) = Area of bar above 19 = 0.1194
Area under curve between 18.5 and 19.5 is 0.4372-0.3176 = 0.1196
Difference = Difference in shaded areas

FIGURE **5.24**
Normal approximation to binomial probabilities for n = 40 and pi = 0.80.

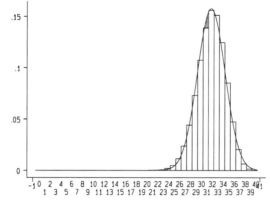

symmetrical about its mean. But now that pi = 0.80, the binomial distribution has become skewed to the left. A visual examination of the difference in the shaded areas shows that for $\pi > 0.50$, we are "adding in" more area than we are "chopping off," so that the approximation is larger than the exact probability. As long as the smaller of $n\pi$ and $n(1-\pi)$ is at least 5, we still do a reasonably good job of approximating binomial probabilities using the normal distribution. Here, the difference in the exact probability and the approximated probability is only $0.1450 - 0.1386 = 0.0064$.

If we change the value of pi to 0.20, we see in Figure 5.25 that we are now "chopping off" more area than we are "adding in," so that we are now underestimating the exact probability by using the normal approximation. The difference in the exact and approximated probabilities is $0.1450 - 0.1513 = -0.0063$.

What happens, then, if the condition that "the smaller of $n\pi$ and $n(1-\pi)$ must be at least 5" is not satisfied? To see this, leave pi = 0.20, and change n to 10. The

FIGURE **5.25**
Normal approximation to binomial probabilities for n = 40 and pi = 0.20.

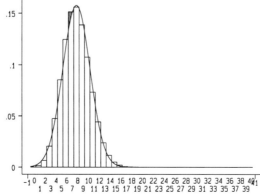

resulting graph, shown in Figure 5.26, tells the story well. It is very easy to see that the "chopped-off" area is much, much larger than the "added-in" area. The difference in the two areas is $0.2285 - 0.2684 = -0.0399$.

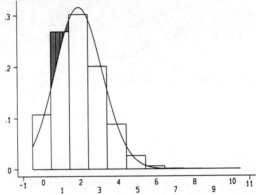

FIGURE 5.26
Normal approximation to binomial probabilities for $n = 10$ and $\mathtt{pi} = 0.20$.

5.5 Summary

In this lengthy guided tour, we have seen a number of different things about sampling from a 0–1 population:

1. *Sampling with replacement:* The distribution of $X =$ number of 1's when drawing a sample of size n with replacement from a population with proportion of 1's equal to π is called a binomial distribution with parameters n and π. This is a discrete distribution, whose mass is centered at $\mu_X = n\pi$. The larger n is, the greater the number of bars in the graph of the distribution. If $\pi = 0.50$, the distribution is symmetrical. If $\pi < 0.50$, the distribution is skewed to the right. If $\pi > 0.50$, the distribution is skewed to the left.

2. *Sampling without replacement:* The distribution of $X =$ number of 1's when drawing a sample of size n without replacement from a population of size N with proportion of 1's equal to π is called a hypergeometric distribution with parameters N, n, and π. This is a discrete distribution, whose mass is centered at $\mu_X = n\pi$. The larger n is, the greater the number of bars in the picture of the distribution. If $\pi = 0.50$, the distribution is symmetrical. If $\pi < 0.50$, the distribution is skewed to the right. If $\pi > 0.50$, the distribution is skewed to the left.

3. *Comparison of binomial and hypergeometric distributions:* The binomial and hypergeometric distributions become more and more alike the larger N is relative to n.

4. *The negative binomial distributions:* If we are interested in the number of draws until the nth 1 is observed when the sample is drawn with replacement from a population with proportion of 1's equal to π, then the distribution of X is

called the negative binomial distribution with parameters n and π. The center of mass (mean) of a negative binomial distribution is n/π. Theoretically, we could continue drawing from our population and never observe the nth 1. How quickly we are likely to observe the nth 1 depends on the values of n and π. If n is large, it will take a large number of draws to observe the nth 1; if n is small, it will likely take a small number of draws. Conversely, if π is small, we are more likely to need a large number of draws; if π is large, the opposite is true.

5. *Approximating binomial probabilities:* Sometimes, calculating exact binomial probabilities can be time-consuming or tedious, or both. In these cases, we can use other distributions to approximate probabilities from the binomial distribution.

   (a) *The Poisson approximation:* If n is large and π is small, the Poisson distribution does well in approximating the binomial distribution. In particular, the larger the value of n and the smaller the value of π, the better the approximation.

   (b) *The normal approximation:* If the smaller of $n\pi$ and $n(1-\pi)$ is at least 5, then the normal distribution does a remarkable job of approximating binomial probabilities. It does especially well when π is close to 0.50 because the binomial distribution, like the normal distribution, is symmetrical for $\pi = 0.50$. As π gets further and further away from 0.50, the larger n needs to be to make up for the lack of symmetry.

5.6 Lab Exercises

5.6.1 Sampling With and Without Replacement

Use the *Sampling With and Without Replacement* Lab to answer the questions in this section.

1. How many bars should appear in a binomial or hypergeometric distribution when n = 15? Run the lab for N = 50, n = 15, and pi = 0.50. Print the resulting graph. Compare the distributions for these values with the distributions for the initial values of N, n, and pi in the lab (N = 50, n = 10, and pi = 0.50). What do you notice about the number of bars? Where are the distributions centered?

2. Let N = 50 and n = 10. Run the lab for each of pi = 0.10, 0.30, 0.50, 0.70, and 0.90. Print each graph. What do you notice about the shapes of the distributions as the value of pi changes? Calculate the mean for each value of pi. Do the changing values of the means mirror the changing shapes of the distributions?

3. Let N = 500. Run the lab and calculate the mean for each of the following values of n and pi. Print all graphs. Describe what you see occurring to the shapes of the distributions in terms of the number of bars, the skewness, and the center of the distribution. How are the changes in the distributions explained by the changes in the values of n and pi?

   (a) n = 4, pi = 0.95
   (b) n = 5, pi = 0.76
   (c) n = 8, pi = 0.475
   (d) n = 10, pi = 0.38
   (e) n = 20, pi = 0.19
   (f) n = 40, pi = 0.095

4 Let pi = 0.50. Print graphs for the following values of N and n. For each graph, calculate the value of the finite population correction factor C. What do you notice about how the changing value of C reflects the changes in the difference between the binomial and hypergeometric distribution?

(a) N = 50, n = 5
(b) N = 50, n = 45
(c) N = 100, n = 5
(d) N = 100, n = 45
(e) N = 300, n = 5
(f) N = 300, n = 45
(g) N = 1000, n = 5
(h) N = 1000, n = 45

5.6.2 The Negative Binomial Distribution

5 Based on the examples in the section discussing the negative binomial distribution, can you describe what the negative binomial distribution will look like for n = 3 and pi = 0.20? Using the *Negative Binomial Distribution* Lab, print the graph for n = 3 and pi = 0.20. Were your conclusions correct?

6 From the *Negative Binomial Distribution* Lab, print graphs for the following values of n and pi. What happens to the overall shape of the distribution as the value of n changes for the same values of pi? What happens to the overall shape of the distribution as the value of pi changes for the same values of n?

(a) n = 1, pi = 0.20, 0.50, and 0.80
(b) n = 3, pi = 0.20, 0.50, and 0.80
(c) n = 7, pi = 0.20, 0.50, and 0.80
(d) n = 10, pi = 0.20, 0.50, and 0.80

5.6.3 Approximating Binomial Probabilities

The Poisson Approximation

7 Run the *Poisson Approximation* Lab for n = 20 and pi = 0.05. Based on the results given in the discussion, do you think the approximation will be better or worse than the approximation for n = 20 and pi = 0.10? At what value of x does the largest difference in the heights of the bars occur? **Optional:** Calculate the difference in the heights of the bars at the mean $n\pi = (20)(0.10)$ and at $x = 0$ using Statistical Tables in StataQuest's Calculator option. Which of the two differences is numerically larger? Is that what you expected to observe? Why?

8 Run the *Poisson Approximation* Lab for pi = 0.05 for each of n = 20, 100, 200, 300, 400, and 500. Print each of the six graphs. What happens to the closeness of the approximation as the value of n increases? **Optional:** Calculate the difference in the heights of the bars at the mean $n\pi$ for each combination. What is happening to the differences in the heights of the bars as n increases?

9 Run the *Poisson Approximation* Lab for n = 100 at each value of pi (0.10, 0.05, 0.04, 0.03, 0.02, and 0.01). Print each of the six graphs. What happens to the closeness of the approximation as the value of pi decreases? **Optional:** Calculate the difference in the heights of the bars at the mean $n\pi$ for each combination. What is happening to the differences in the heights of the bars as pi decreases?

10 Run the *Poisson Approximation* Lab for the following values of n and pi, and then print out each of the graphs. What is happening to the closeness of the two distributions as n increases? What about as pi decreases?

(a) n = 100, pi = 0.10, 0.05, and 0.01
(b) n = 300, pi = 0.10, 0.05, and 0.01
(c) n = 500, pi = 0.10, 0.05, and 0.01

Optional: Calculate the difference in the heights of the bars at the mean $n\pi$ for each combination. As n increases, what is happening to the difference in the heights of the bars? What is happening to the difference in the heights of the bars as pi decreases?

The Normal Approximation

11 Run the *Normal Approximation* Lab for n = 15, 20, and 50 with pi = 0.50. Print the three graphs. What happens to the closeness of the exact probabilities and the approximated probabilities as the value of n increases?

12 Run the *Normal Approximation* Lab for n = 30 with pi = 0.20, 0.50, and 0.80. Print the three graphs. What happens to the closeness of the exact probabilities to the approximated probabilities as the value of pi changes?

13 Run the *Normal Approximation* Lab for n = 20 with pi = 0.10, 0.20, 0.50, 0.80, and 0.90. Print all five graphs. Comment on the appropriateness of using each of the five combinations, and explain what happens to the closeness of the exact probabilities to the approximated probabilities as the value of pi changes. What would be an alternative to using the normal approximation when pi = 0.10? Compare this alternative with the normal approximation. Which is better, the alternative or the normal approximation?

5.7 Formulas for Discrete Distributions, Means, and Variances

Distribution	Mean	Variance
Binomial with parameters n and π: Arises in either (1) the number of 1's when sampling (with replacement) from a 0–1 population where the proportion of 1's in the population is π, or (2) the number of "successes" in n independent "trials," where each trial can be either a success or a failure and the probability of a success on one trial is π.		
$\binom{n}{x}\pi^x(1-\pi)^{n-x}$ $\quad x = 0, 1, \ldots, n, \quad 0 \leq \pi \leq 1$	$n\pi$	$n\pi(1-\pi)$

Hypergeometric with parameters N, n, and M: The number of 1's in a sample (without replacement) of size n from a 0–1 population of size N, where the number of 1's in the population is M (and thus the proportion of 1's in the population is $\pi = M/N$).

$$\frac{\binom{M}{x}\binom{N-M}{n-x}}{\binom{N}{n}} \qquad n\pi \qquad \left(\frac{N-n}{N-1}\right)n\pi(1-\pi)$$

$$\max(0, M - (N - n)) \leq x \leq \min(n, M)$$

Negative binomial with parameters n and π: The number of the binomial trial where the nth success occurs when on each trial the probability of success is π.

$$\binom{x-1}{n-1}\pi^n(1-\pi)^{x-n} \qquad \frac{n}{\pi} \qquad \frac{n(1-\pi)}{\pi^2}$$

$$x = n, n+1, n+2, \ldots$$

Poisson with parameter λ: Arises in two ways: (1) If we have a series of events in which the time between events has the exponential distribution with mean μ, then the number of events in a time interval of length T has the Poisson distribution with $\lambda = T/\mu$. (2) If the binomial parameter n is large and π is small, then the binomial distribution can be approximated by the Poisson distribution with $\lambda = n\pi$.

$$\frac{\lambda^x e^{-\lambda}}{x!} \qquad \lambda \qquad \lambda$$

$$x = 0, 1, 2, \ldots, \quad \lambda > 0$$

6

Bivariate Descriptive Statistics

In the *Random Sampling* and *How Are Populations Distributed?* Labs, we considered univariate data—that is, each element in a population or sample had only one numerical characteristic. In many real-world data sets, two or more variables are measured on each element in what is called a *bivariate data set*.

6.1 Introduction

6.1.1 Two Sampling Schemes for Bivariate Populations

Suppose we want to study the relationship of two characteristics (such as height and weight or SAT and GPR) for a large population of objects (such as the students at a large university), and we believe that one of the characteristics (denoted by Y) is being influenced by the other one (denoted by X). For example, it is commonly believed that people's weights are a function of their heights, and not vice versa, or that their GPR is a function of their SAT. There are two ways we can select a sample of size n from the population, leading to the data denoted by $(X_1, Y_1), (X_2, Y_2), \ldots, (X_n, Y_n)$:

1. *Random X's:* In this case, we simply select a random sample of objects and determine their Y's and X's.
2. *Fixed X's:* In this case, we specify before doing the sampling what values of X we want to have (and how many of each). Then, for each value of X, we randomly select the desired number of objects having that X and determine the value of Y for each one of the objects. We might do this if we were studying height and

weight and wanted to ensure we had people with a wide range of heights (X's) in our sample.

In some instances, we have no preconceived notion of one of the two characteristics being a function of the other.

6.1.2 Scatterplots

With a bivariate data set, the first step typically is to draw a scatterplot of the data. In Figure 6.1, we have drawn two scatterplots of the heights (X) and weights (Y) of 18 people. The top graph in the figure shows how the plot is formed; for each person, a symbol is drawn (in this case, a circle) at the point where a vertical line drawn above the height and a horizontal line drawn to the right of the weight would intersect. The bottom graph places the value of the actual data at each point.

We can see two main features in this plot:

1. As the height of people increases (moving from left to right on the horizontal axis), their weight tends also to increase. This relationship appears to be very strong. If two variables are not strongly related, the points will fall more randomly and uniformly in the plot.

2. The points on the plot appear to be going up linearly, that is, seem to be scattered around a line having positive slope.

The next two sections try to quantify these two ideas in terms of the correlation coefficient and the least squares regression line. The *correlation coefficient* is a number that tries to measure the direction and strength of a linear relationship. The *least squares regression line* is an attempt to draw a line on the graph that "best fits" a scatterplot of points.

6.1.3 Correlation Coefficient

We seek a number that measures both how closely linearly related two variables are and what direction (positive or negative) the relationship goes in. The correlation coefficient (for a sample denoted by r and for a population denoted by ρ) does just that. It is a number that ranges from -1 to $+1$ and for which the following apply:

1. The closer r or ρ is to ± 1, the stronger the linear relationship.
2. The closer r or ρ is to 0, the weaker the linear relationship. If ρ is actually 0, we say that X and Y are not linearly related, which is also called *uncorrelated*.
3. The sign of r or ρ tells whether the relationship is positive or negative (going up or down as X goes up).

The correlation coefficient r is defined as follows (ρ is defined in the same way, except for a whole population instead of simply a sample). We start by calculating the sample means $\bar{X}$ and $\bar{Y}$ of the set of X's and Y's. Then we find the *deviations* $X_i - \bar{X}$ and $Y_i - \bar{Y}$ for each (X_i, Y_i) point in our data set. The following features of these deviations are what make r (defined in a moment) meaningful:

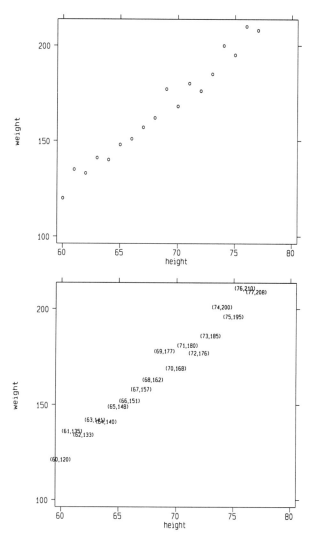

FIGURE **6.1**
Two versions of a scatterplot. The top graph puts circles at each point; the bottom one places the value of the data at each point.

1. A deviation is positive (negative, zero) if the value of the variable is above (below, equal to) the average of the values of the variable.

2. The product $(X_i - \bar{X})(Y_i - \bar{Y})$ of the X and Y deviations for one point in the data set is positive if the two individual deviations $(X_i - \bar{X})$ and $(Y_i - \bar{Y})$ are either both positive or both negative. On the other hand, the product is negative if the individual deviations have opposite signs (one positive, one negative).

Thus, if we were to add the product of the deviations together and the result were positive, this would indicate that the products of deviations tended to have the same sign. That is, large X's tend to be associated with large Y's (both deviations positive), small X's tend to be associated with small Y's (both deviations negative), and so on. This means the relationship is positive. On the other hand, if the sum of products of deviations is negative, then large X's tend to be associated with small Y's, and small X's tend to be associated with large Y's (this gives the individual deviations opposite signs).

Now we can finally define r:

$$r = \frac{\sum(X_i - \bar{X})(Y_i - \bar{Y})}{(n-1)s_X s_Y},$$

where s_X and s_Y are the sample standard deviations of the X's and Y's. Thus, the numerator of r is the sum of the products of deviations that we have been talking about. Because standard deviations such as s_X and s_Y can never be negative, the sign of r is the same as the sign of its numerator.

The effect of dividing by $(n-1)s_X s_Y$ is twofold:

1. r falls between -1 and $+1$, with ± 1 possible only if the points on the scatterplot all fall exactly on a line.

2. The same value of r is obtained no matter what units the X and Y variables are measured in. For example, if we calculated r for heights and weights measured in inches and pounds, converted the data to centimeters and kilograms, and then calculated r again for the new data, we would get exactly the same value!

6.1.4 Least Squares Regression Line

Given a set of points such as the one in Figure 6.1, we seek to find a line that "best fits" the points. The line that is most often used (called the least squares regression line for reasons that will become apparent) is obtained as follows.

If we draw any particular line on the plot, we can find the *vertical deviation* of each point to the line; that is, for each point, we can subtract the Y value of the line from the Y value of the point. If this is positive (negative, zero), then the point is above (below, on) the line. Any reasonable line will have some points above it and some below. If we take the absolute value of the vertical deviations we get the vertical *distances* of the points to the line. The so-called least squares regression line is that line whose sum of squares of vertical distances is smaller than that of any other possible line. Figure 6.7 later in the chapter shows a scatterplot of a random sample of points from a population, as well as the least squares regression lines for the sample (the top line) and for the entire population (the bottom line). In this figure, we have placed vertical line segments from each point in the sample to the sample least squares regression line. The lengths of these line segments are the vertical distances. Any possible line will have its own set of these distances, and the least squares regression line will be the one whose set of vertical distances is the smallest.

A miracle of mathematics is that we don't have to try all possible lines to determine which one is the best one. Rather, with a little algebra or calculus, we can prove that the best line is the one whose slope and intercept are given by

$$b = \frac{\sum(X_i - \bar{X})(Y_i - \bar{Y})}{\sum(X_i - \bar{X})^2}, \quad a = \bar{Y} - b\bar{X}.$$

Thus, given a set of data, it is very easy to find the least squares regression line as the one having this slope and this intercept. To draw the line on a scatterplot, we need only pick an X at each end of the horizontal axis, calculate $Y = a + bX$ for each of the two X's, plot the two resulting (X, Y) points on the graph, and connect them with a line.

6.1.5 Residuals and the Multiple Correlation Coefficient, r^2

The vertical deviations of the points from the least squares regression line are often denoted by $e_1, e_2, \ldots, e_n$ and play an important role in analyzing such (X, Y) data:

1. If we denote by $\hat{Y}_i$ the Y value on the line at the ith data point, then $\hat{Y}_i$ is the value that the line says Y_i should have been. Thus, it is called a "predicted" or "fitted" value.
2. $e_i = Y_i - \hat{Y}_i$. Thus, the e's are called *residuals*, because they are the difference between the observed values of the Y's and the predicted values.
3. The residuals add to zero.
4. The sum of squares of residuals is denoted by RSS and is a measure of how close the points are to the line.

Although RSS is a measure of how close the points are to the line, it is not independent of the units of measure of the data. A measure that does have this property is called the *multiple correlation coefficient*. This measure is usually denoted by the symbol r^2 and can be calculated by

$$r^2 = 1 - \frac{s_e^2}{s_Y^2},$$

where s_e^2 and s_Y^2 are the sample variances of the residuals and the observed Y's, respectively.

The multiple correlation coefficient has the following properties:

1. r^2 is the square of the correlation coefficient r between the X's and Y's.
2. r^2 is always between 0 and 1. It has the value 1 if and only if the points all fall exactly on the line.
3. Thus, $100r^2$ is a percentage and can be interpreted as the percentage of the variability in the Y's that is "explained" by the relationship between Y and X.

This last point is very important. For our height and weight example, we would like to explain why people's weights (Y) are not all the same. One obvious factor leading to variability in weight is the variability in people's heights (X) and the fact that people's weights are a linear function of their heights. But how much of the variability in Y is due to variability in X? If 100% of the variability were because of X, the points would fall perfectly on the line, and the residuals would all be 0. Thus, s_e^2 would be 0 and r^2 would be 1!

Conversely, if none of the variability in Y were due to its relationship with X, then the points would fall randomly and uniformly around the least squares regression line, and the residuals would basically vary the same amount as the original Y's did. Thus, s_e^2 and s_Y^2 would be basically the same size, their ratio would be close to 1, and r^2 would be close to 0.

6.2 Objectives

1. Draw a scatterplot for a sample from a bivariate population.
2. Examine what the sample correlation coefficient says about the scatterplot.
3. Determine the least squares line for a scatterplot.
4. Use r^2 to measure how well the least squares line fits the data.

6.3 Description

To begin the lab, select Bivariate Descriptive Statistics from the Labs menu. The following submenu appears with a choice of three concept labs:

> Scatterplots I
> Scatterplots II
> Least Squares

6.3.1 Scatterplots I

Choosing this concept lab opens a graphics window that contains scatterplots of samples of size $n = 500$ from bivariate populations having 20 different values of the population correlation coefficient ρ. The values of ρ range from -0.9 to 0.9 in increments of 0.1, with two samples for $\rho = 0.0$. See Figure 6.2 for an example.

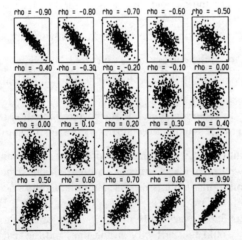

FIGURE 6.2
Example of the *Scatterplots I* Lab.

The dialog box shown in Figure 6.3 briefly summarizes what the lab does and what you should notice about the scatterplots. In the bottom left-hand corner of the dialog box are boxes (labeled n and Symbol #) where you can choose which value of n to use and which symbol to use to represent the points in the scatterplots. The

FIGURE 6.3
Dialog box for the
Scatterplots I Lab.

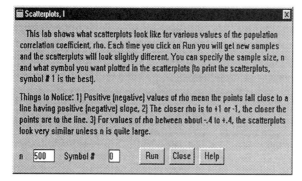

initial values of these items are 500 and 0. The value of n must be at least 2 and at most 600, while Symbol # must be 0, 1, 2, 3, 4, 5, or 6 (0 gives a single pixel at each point, 1 a circle, 2 a square, 3 a triangle, 4 a circle, 5 a diamond, and 6 a plus sign).

To the right of the two boxes are three buttons:

1   Run: Clicking on this button causes the lab to produce another set of scatterplots with the values of n and Symbol # that have been entered in the dialog box.
2   Close: Clicking on this button closes the lab and returns you to the StatConcepts menus.
3   Help: Clicking on this button opens a help window containing information about the lab.

6.3.2 Scatterplots II

Choosing this concept lab opens a graphics window that contains scatterplots of samples of size $n = 100$ from bivariate populations having $\rho = 0.0$, $\rho = 0.9$, $\rho = -0.9$, and $\rho = 0.99$. See Figure 6.4 for an example.

FIGURE 6.4
Example of the
Scatterplots II Lab.

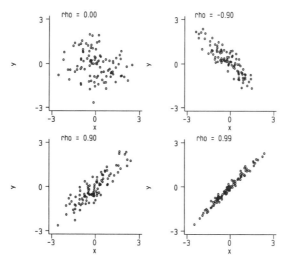

The dialog box shown in Figure 6.5 briefly summarizes what the lab does and what you should notice about the scatterplots. In the bottom left-hand corner of the dialog box are boxes (labeled n and Rho) where you can choose which values of *n* and ρ to use and which symbol to use to represent the points in the scatterplot. The initial values of these items are 100 and 0.99. The value of n must be at least 2 and at most 600, while Rho can be any number greater than -1 and less than $+1$.

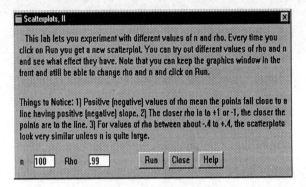

FIGURE **6.5**
Dialog box for the *Scatterplots II* Lab.

To the right of the two dialog boxes are three buttons:

1. Run: Clicking on this button causes the lab to produce another scatterplot with the values of n and Rho that have been entered in the dialog box. The first time you click on Run, the screen will be cleared before producing the scatterplot. On subsequent clicks, the scatterplot will move to the lower left of the window, then to the upper right, and then the lower right. On the next click, the window will clear and the scatterplot will move to the upper left, and the process will begin again. Each time you click on Run, the scatterplot will use the chosen values of n and Rho.

2. Close: Clicking on this button closes the lab and returns you to the StatConcepts menus.

3. Help: Clicking on this button opens a help window containing information about the lab.

6.3.3 Least Squares

Choosing this concept lab brings up the dialog box shown in Figure 6.6, which briefly summarizes what the lab does and what you should notice during the lab.

In the bottom left-hand corner of the dialog box are three buttons:

1. Run: Clicking on this button causes the lab to produce either a graph similar to the one shown in Figure 6.7 if # of Samples is 1, or a graph similar to the one in Figure 6.8 if # of Samples is more than 1. (We will develop this issue when we discuss the values under # of Samples.)

2. Close: Clicking on this button closes the lab and returns you to the StatConcepts menus.

FIGURE **6.6**
Dialog box for the *Least Squares* Lab.

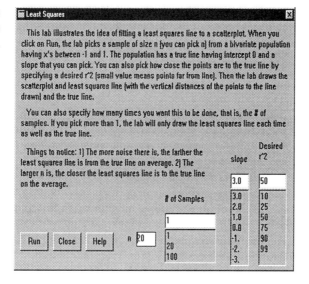

FIGURE **6.7**
Example of the *Least Squares* Lab for `# of Samples = 1`.

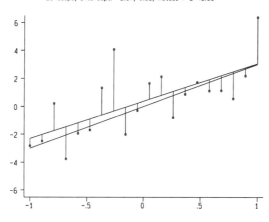

FIGURE **6.8**
Example of the *Least Squares* Lab for `# of Samples = 20`.

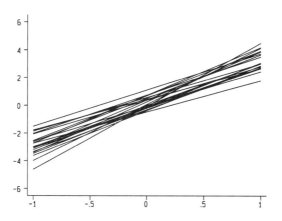

3 `Help`: Clicking on this button opens a help window containing information about the lab.

To the right of these buttons are four boxes labeled `n`, `# of Samples`, `slope`, and `Desired r`^{2}:

1 `n`: The value here specifies the size of the sample or samples to be used in the lab.
2 `# of Samples`: This can be either 1 or something greater than 1:
   (a) If the value is 1, then clicking on `Run` will result in the selection of a random sample of size n from a bivariate population whose true regression line has intercept 0, slope as specified in the `slope` item, and true value of r^2 (expressed as a percentage) as specified in the `Desired r`^{2} item. The scatterplot of the sample is then drawn, as are the least squares regression lines for the sample and the population (the sample line in yellow, the population line in red). Line segments representing the vertical deviations are included. Finally, the slope, intercept, and r^2 for the sample are displayed at the top of the graph.
   (b) If the value is more than 1, then a series of samples are drawn, with the population least squares regression lines plotted in red and the lines for each sample plotted in yellow.
3 `slope`: This item allows you to specify the value of the slope of the population line.
4 `Desired r`^{2}: This item allows you to specify the value of r^2 for the population.

The initial values for the four items are n = 20, `# of Samples` = 1, `slope` = 3.0, and `Desired r`^{2} = 50%.

Note that the samples that are taken are for *fixed X*'s that are n equally spaced from −1 to +1 and that all graphs produced during the lab are on the same scale of X from −1 to +1 and Y from −6 to +6. This allows the results in any graphs to be comparable.

6.4 Guided Tour of the Lab

6.4.1 Scatterplots I

We begin our tour of this lab by studing what scatterplots of samples from populations having varying values of ρ look like. Begin the lab by clicking on `Scatterplots I` from the `Bivariate Descriptive Statistics` submenu. You will see a set of graphs like those shown in Figure 6.2. Each scatterplot is of a random sample of 500 observations from a population, with the value of ρ displayed above the scatterplot. Notice the following from the graphs:

1 All of the scatterplots for negative values of ρ tend to fall around a line of negative slope, and as ρ increases from −0.9 to −0.1, the scatter gets less and less concentrated around a line.
2 The same sort of phenomenon occurs for positive values of ρ except that the linear relationship has a positive slope.

3. Even for samples this large, it is very difficult to tell the difference among scatterplots for nearby values of ρ. For example, those plots for ρ between -0.2 and 0.2 look very similar.

4. Although the values of ρ in the plots are increasing by the same amount from one plot to the next, the appearance of the scatter does not seem to change by the same amount.

Now change the value of n to 50 and the Symbol # to 1 and click on Run. The result will look something like the scatterplots shown in Figure 6.9. The main point of this figure is to illustrate that with samples of this size, it is even harder to tell visually the value of ρ. Click several more times on Run and see how much variability there is from one set of scatterplots to another.

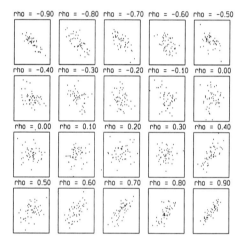

FIGURE 6.9 Scatterplots for samples of size 50.

6.4.2 Scatterplots II

In the tour of this lab, we will look at scatterplots for a wide variety of values of the correlation coefficient and sample size so that you can get a feel for what the correlation coefficient tells us.

Select Scatterplots II from the Bivariate Descriptive Statistics submenu. You will see a set of graphs like those shown in Figure 6.4. The relationship for the positive (negative) values of ρ has Y increasing (decreasing) as X increases, while for $\rho = 0$ it is hard to discern any linear pattern in the scatterplot at all.

Now change n to 50 and Rho to 0.7 and click on Run four times. You should see something like the top set of plots shown in Figure 6.10. This is what scatterplots for samples of size 50 look like from a population with $\rho = 0.7$. Click on Run several more times to be sure you have a feel for what they look like. Now change ρ to -0.7 and click on Run several times, which should result in something like the bottom set of plots shown in Figure 6.10. That is, the degree of scatter is the same, but the direction of the relationship is now negative instead of positive. You should do this for several combinations of ρ and n to be sure you know what scatterplots mean.

FIGURE 6.10
Four scatterplots for n = 50 and Rho = 0.7 (top) and Rho = −0.7 (bottom).

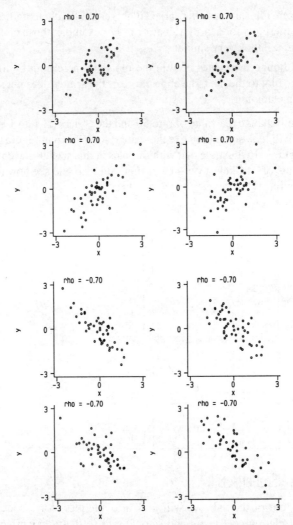

Next, we'll produce a graph like the one shown in Figure 6.11, namely, scatterplots for Rho = 0.9 and n = 50, 100, 150, and 200. Start the lab again, change Rho to 0.9 and n to 50, and then click on Run. Then change n to 100, 150, and 200, clicking on Run each time.

6.4.3 Least Squares

We begin our tour by selecting Least Squares from the Bivariate Descriptive Statistics submenu. You should read the information in the dialog box and review the material in the preceding sections of this chapter before doing the lab.

Note the initial values of the items in the dialog box: Sample size n = 20, # of Samples = 1, the slope of the population line is 3, and the Desired r^2 is 50%.

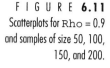

FIGURE **6.11**
Scatterplots for Rho = 0.9 and samples of size 50, 100, 150, and 200.

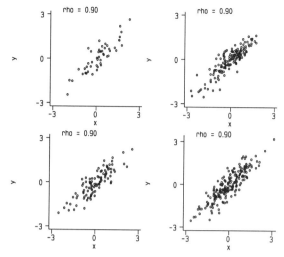

Using these values, click on Run, and you will get a plot similar to the one shown in Figure 6.7.

What Does the Slope Mean?

The true line is drawn in red in this plot. For $X = -1$, $Y = -3$, while at the other end of the line, $X = 1$ and $Y = 3$. Thus, Y increases by 6 as X increases by 2. This verifies that the slope of the line is, in fact, $3 = 6/2$. This is a characteristic of a line. If you increase X by one unit, then Y increases by an amount equal to the value of the slope. Now change the value of slope to -3 and click on Run again. The resulting graph will look similar to the one shown in Figure 6.12 and very similar to the one shown in Figure 6.7, except that the line goes down as much as it went up when the slope was 3. Now Y goes down 6 units as X increases from -1 to 1. Note also that

FIGURE **6.12**
One sample for slope = -3.

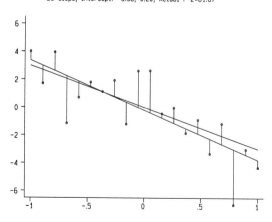

the farther the slope is from 0, the steeper the true line becomes. You should try other values of the slope and see what effect this has on the line.

What Does r^2 Mean?

Change slope back to 3 and click on Run. Visualize how much scatter there is of the points around the line and how far the line for the sample (the yellow one) is from the line for the population. Click on Run several more times so you can get a real feel for what a typical scatterplot looks like for this situation. Note that the population r^2 is 50%, which is halfway between 0% and 100%. Recall that 100% would mean that the points all fall right on the line. The Desired r^2 item in the dialog box lets you choose values of 10, 25, 50, 75, 90, and 99%. Choose 25% and click on Run. You will get a plot similar to the top one shown in Figure 6.13. Notice how the points

FIGURE **6.13**
One sample for r^2 of 25% (top) and 75% (bottom) with n = 20.

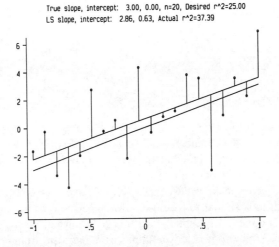

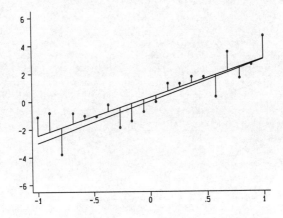

tend to be more scattered around the yellow line than for the 50% case. Click on Run several more times to see how the scatter looks and how far the sample line is from the population line. We can study this distance in more detail by changing the # of Samples item in the dialog box to 20 and clicking on Run. The resulting plot will look like the top one shown in Figure 6.14. Compare this to Figure 6.8, which is the same type of plot except for r^2 being 50%. Notice how the lines are much more variable for the smaller r^2.

Now let's do the same thing for r^2 of 75%. Change the Desired r^2 value in the dialog box to 75% and the # of Samples back to 1, and click on Run. Now there is much less scatter than in the 25% case. The bottom plot in Figure 6.13 is an example of what you should see. Click on Run several more times to see a typical amount of scatter for r^2 of 75%. Then change # of Samples back to 20 and click on Run. Notice how much more tightly grouped the 20 sample lines are for this r^2

FIGURE **6.14**
Twenty samples for r^2 of 25% (top) and 75% (bottom) with $n = 20$.

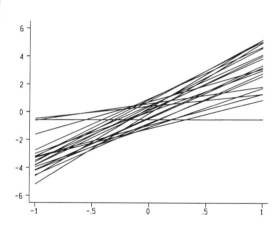

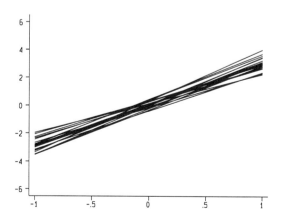

than for the 50% case (Figure 6.8) and in particular for the 25% case (the bottom plot in Figure 6.14).

What Is the Effect of Sample Size?

In the previous two parts of this tour, we did everything for sample size n of 20. Now we'll increase it to 100 and see what effect this has. Start with `Desired r`^{2} of 50% and n = 100, and click on `Run`. Then compare the result, shown in Figure 6.15, with Figure 6.7, which has the same conditions except for smaller sample size. The points still seem to be scattered approximately the same amount (the value of r^2 is 50% in both cases), but the sample line will tend to be closer to the average with our larger sample size. To see this better, change `# of Samples` to 20, click on `Run`, and compare the result, shown in Figure 6.16, with Figure 6.8. Notice how the lines are grouped more closely together when we have a larger sample size.

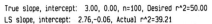
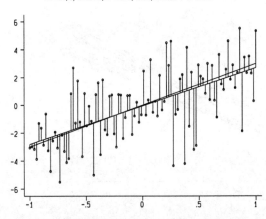

FIGURE **6.15** One sample for r^2 of 50% and sample size 100.

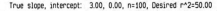
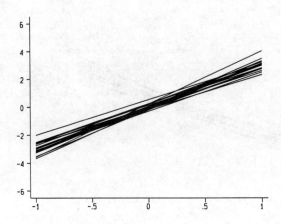

FIGURE **6.16** Twenty samples for r^2 of 50% and sample size 100.

6.5 Summary

We have seen the following ideas in action in this chapter:

1. A scatterplot of a sample can tell us a lot about a population.
2. The sample correlation coefficient measures the strength of the linear relationship between two variables. It has values between -1 and $+1$, where the sign tells the direction (positive or negative) of the relationship and the closeness of the value to ± 1 tells how strong the relationship is.
3. The least squares regression line is the line that best fits a scatter of points in the sense of minimizing the sum of squares of vertical distances of the points to the line.
4. Residuals play an important role in describing the relationship between X and Y.
5. The multiple correlation coefficient r^2 tells how well the line fits a set of points and how much of the variability in Y is explained by its linear relationship with X.
6. Sample size and r^2 play an important role in how well the least squares regression line for a sample agrees with the one for the whole population.

6.6 Lab Exercises

6.6.1 Scatterplots I

1. Produce plots similar to those in Figure 6.9, but substitute these values:
   - (a) n = 100
   - (b) n = 200
   - (c) n = 300
   - (d) n = 400
2. Using sample size 50 and circles as the plotting symbol, run the *Scatterplots I* Lab. Which of the two plots for $\rho = 0$ looks more "random," that is, uncorrelated?

6.6.2 Scatterplots II

3. Following the first part of the *Scatterplots II* guided tour, produce plots similar to those in Figure 6.10, but substitute these values:
   - (a) n = 50, Rho = 0.9
   - (b) n = 50, Rho = -0.9
   - (c) n = 50, Rho = 0.0
   - (d) n = 100, Rho = 0.9
   - (e) n = 100, Rho = -0.9
   - (f) n = 100, Rho = 0.0
4. Following the second part of the *Scatterplots II* guided tour, produce plots similar to those in Figure 6.11, but substitute these values:
   - (a) Rho = -0.9, n = 50, 100, 150, and 200
   - (b) Rho = 0.2, n = 50, 100, 150, and 200
   - (c) Rho = -0.2, n = 50, 100, 150, and 200
   - (d) Rho = 0.5, n = 50, 100, 150, and 200

100 Chapter 6 Bivariate Descriptive Statistics

(e) Rho = −0.5, n = 50, 100, 150, and 200

(f) Rho = 0.7, n = 50, 100, 150, and 200

(g) Rho = 0.8, n = 50, 100, 150, and 200

5 Do the last two parts of the previous exercise, and then comment on whether you can tell scatterplots for $\rho = 0.7$ and $\rho = 0.8$ apart just by looking. Is it easier to tell them apart for larger sample sizes?

6.6.3 Least Squares

6 Following the guided tour, produce plots similar to Figure 6.13, but substitute these values:

(a) n = 40

(b) n = 60

(c) n = 80

(d) n = 100

7 Following the guided tour, produce plots similar to Figure 6.14, but substitute these values:

(a) n = 40

(b) n = 60

(c) n = 80

(d) n = 100

8 Following the guided tour, produce plots similar to Figure 6.15, but substitute these values:

(a) Desired r^2 = 10%

(b) Desired r^2 = 25%

(c) Desired r^2 = 75%

(d) Desired r^2 = 99%

7

Central Limit Theorem

In the *Random Sampling* Lab, we got a sneak preview of the tour we will take to learn more about one of the most important tools in statistics, the *Central Limit Theorem*.

7.1 Introduction

7.1.1 Some Basic Ideas

In the *Random Sampling* Lab, we saw that the value of the sample mean changes depending on which elements are chosen for the sample. The sampling distribution of the sample mean is defined as the distribution that describes the pattern of all the different values we could possibly observe for the sample mean. The Central Limit Theorem promises us that no matter what distribution a sample is chosen from, if the sample size is large enough, the sampling distribution of the sample mean will be a normal distribution!

The mean of the sampling distribution of $\bar{X}$ is the mean μ of the parent population. This implies that the sample mean $\bar{X}$ is an *unbiased estimator* of the population mean μ. The values of the sample mean give accurate estimates of the population mean, that is, they are "on target."

The variance of the sampling distribution of $\bar{X}$ is σ^2/n, where σ^2 is the variance of the parent population. Because the variance of the sampling distribution is divided by the sample size n, if the sample size increases, the variability of the values we could observe for the sample mean decreases. In other words, the larger our sample size becomes, the more precise $\bar{X}$ becomes at estimating the population mean μ.

Thus, not only are we "on target," but we are hitting closer to the target (μ) more *consistently* as the sample size increases.

One property of the normal distribution is that approximately 95% of the values X in the distribution will fall in the interval $\mu_X \pm 2\sigma_X$. This result is sometimes referred to as the *Empirical Rule*. The values we are studying are the $\bar{X}$'s from many different samples; the mean of $\bar{X}$ is μ, and the standard deviation of $\bar{X}$ is $\sigma/\sqrt{n}$. Therefore, this interval becomes $\mu \pm 2\sigma/\sqrt{n}$. We will use this result to determine if means calculated from samples of size n from non-normal populations are normally distributed.

7.2 Objectives

1. Illustrate that the shape of the distribution of many sample means, each calculated from a different, randomly selected sample, is a bell-shaped, or normal, curve.

2. Illustrate that the mean of the sampling distribution of the sample mean is the mean of the parent distribution (the mean of the means is the mean).

3. Examine how the larger the sample size gets, the more bell-shaped the sampling distribution of the sample means becomes.

4. Examine how the larger the sample size, the narrower the shape of the sampling distribution of the sample means.

5. Examine how the more skewed the parent distribution is, the larger the sample size n needs to be before the sampling distribution becomes normal.

7.3 Description

Begin the lab by clicking on Central Limit Theorem under the Labs menu. This will open the dialog box shown in Figure 7.1, which briefly summarizes the lab. In the bottom left-hand corner of the dialog box are three buttons:

1. Run: Clicking on this button runs the lab for the choices designated in the boxes to the right. The results of clicking on Run will differ, depending on what is specified in the Choice box. This will be described in more detail when the options under Choice are discussed.

2. Close: Clicking on this button closes the lab and returns you to the StatConcepts menus.

3. Help: Clicking on this button opens a help window containing information about the lab.

To the right of these buttons are three boxes labeled n, Sample from, and Choice. The values specified in the box labeled Choice will determine which type of experiment you conduct when you click on Run.

1. n: The value specified in this box determines the size of the samples that will be chosen to calculate the sample means. The value of n is not restricted to those shown. If you wish the sample size to be something other than what is shown, simply type the desired sample size in the box. The initial value of n is 10.

FIGURE 7.1
Dialog box for *Central Limit Theorem* Lab.

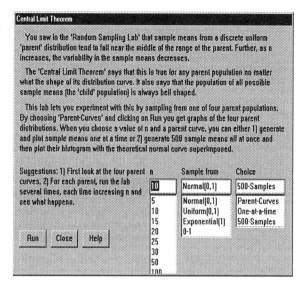

2 Sample from: The distribution specified in this box is the distribution of the population from which samples are to be chosen. The distributions are restricted to those choices given. When Choice is Parent-Curves, graphs of these distributions are displayed. The initial distribution in Sample from is Normal(0,1).

3 Choice: The options here determine which experiment is to be conducted when the Run button is clicked. The initial Choice is 500-Samples. The three choices and their outcomes are given below:

(a) Parent-Curves: If this is the specified choice, clicking on Run opens a graphics window containing four graphs each of a different parent distribution (see Figure 7.2). These parent distributions are the choices specified under Sample from.

FIGURE 7.2
Four parent distributions in the *Central Limit Theorem* Lab.

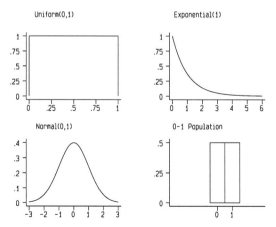

(b) One-at-a-time: If this choice is specified, clicking on Run will cause the lab to randomly select a sample of size n from the distribution specified in Sample from, calculate the sample mean for that sample, and draw a small yellow box directly above the value on the horizontal axis closest in value to the sample mean. The process is repeated until the stack of means becomes too tall to fit on the vertical axis (see Figure 7.3). All graphs will differ somewhat in appearance because a different set of samples is selected anytime the lab is run.

(c) 500-Samples: If this choice is specified, StatConcepts will randomly select 500 samples of size n from the parent distribution specified in Sample from. For each of the 500 samples, the sample mean is calculated. Then a graphics window opens containing a histogram of the 500 means with the corresponding normal distribution in red superimposed on the histogram (see Figure 7.4).

FIGURE 7.3
One-at-a-time graph of the distribution of means for samples with $n = 10$ from a $N(0,1)$ population.

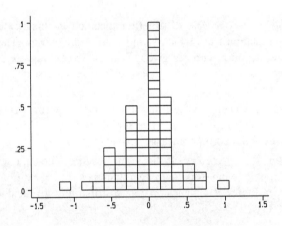

FIGURE 7.4
Normal distribution superimposed on a histogram of 500 means of samples with $n = 10$ from a $N(0,1)$ population.

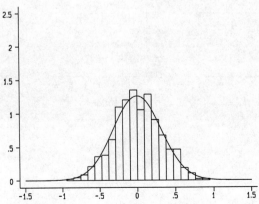

7.4 Guided Tour of the Lab

Under Choices, there are three options. Our tour takes us through the properties and behavior of the sampling distribution of $\bar{X}$ for each of the three choices. At the end of the tour, we will have witnessed something truly remarkable—no matter what the population distribution is, if the sample size n is large enough, the sampling distribution of the sample mean is a normal distribution!

7.4.1 Parent-Curves

We begin our tour by examining the four population distributions from which samples in this lab are chosen. Recall that a population is a collection of elements. Each element has some characteristic or property that we are interested in and that can be measured. A *parent distribution* is the function that describes the shape, or pattern, of how likely each measurement is to occur. Population distributions are described in detail in the *How Are Populations Distributed?* Lab, where it is shown that they take many shapes and forms but that the total area under any curve is exactly equal to 1. In a graph of a distribution, the horizontal axis represents values of the measurements, while the vertical axis represents the density of those values in the population. Therefore, the height of the curve of a particular distribution tells us how tightly clustered the elements of a population are around a particular value on the horizontal axis.

Open the lab, chose Parent-Curves under Choice, and click on Run. A graphics window will open containing the four graphs shown in Figure 7.2. Each of these graphs shows one of the parent distributions that may be selected in Sample from. They are the population distributions from which samples are chosen. Here, we briefly discuss each of the distributions and some of their important properties. The *How Are Populations Distributed?* Lab contains a more complete discussion of the uniform, exponential, and normal distributions. The 0–1 population is discussed in the *Sampling From 0–1 Populations* Lab.

The Uniform(0,1) Distribution

The Uniform(0,1) distribution is shown in the upper left quadrant of Figure 7.2. Looking at this graph, it is easy to see why this distribution is called "uniform." Elements in a population with a uniform distribution are evenly (or uniformly) spread out over the entire interval. In this particular example of a uniform distribution, the measured values of the characteristic fall between 0 and 1—hence, the *Uniform(0,1)* distribution. The mean of this distribution is $\mu = 0.5$, and the variance is $\sigma^2 = 1/12$. Notice that the uniform distribution is symmetrical about its mean.

The Exponential(1) Distribution

The Exponential(1) distribution is shown in the upper right quadrant of Figure 7.2. This distribution is severely skewed to the right. Practically speaking, this means that the population contains more elements with small values than elements with large values. The Exponential(1) distribution has mean $\mu = 1$ and variance $\sigma^2 = 1$.

The Normal(0,1) Distribution

The Normal(0,1), or *standard normal*, distribution is shown in the bottom left quadrant in Figure 7.2. The mean of the standard normal distribution is $\mu = 0$, and the variance is $\sigma^2 = 1$. Notice that like the Uniform(0,1) distribution, the normal distribution is symmetrical around its mean. But for the normal distribution, the farther away we get from the mean in either direction, the less likely it is that the value will occur in the population. The values in a normal population cluster most tightly around the population mean. Contrast this property with the uniform distribution, where all values are equally likely.

The 0-1 Population

The 0-1 distribution is shown in the bottom right quadrant of Figure 7.2. A 0–1 population is a set of elements for which the characteristic measured can take on only two values, coded as either 0 or 1. In a 0–1 population, the proportion π of 1's is also the mean of the population. The 0–1 population considered in this lab has a proportion of 1's $\pi = 0.5$. The variance of this 0–1 population is $\sigma^2 = \pi(1 - \pi) = (0.5)(0.5) = 0.25$. Because $\pi = 0.5$, this 0–1 population is symmetrical about its mean.

7.4.2 One-at-a-time

Specify One-at-a-time under Choice. Run the lab for samples of size n = 10 from a Normal(0,1) distribution. Because the population distribution is normal, the sampling distribution of $\bar{X}$ is also normal, regardless of the sample size. This case will be a good control group, or baseline, to use for comparison purposes when the population distribution is not normal. The resulting graph will look somewhat like the one shown in Figure 7.3. To produce this graph, StatConcepts generates a sample of size $n = 10$ from a Normal(0,1) distribution, calculates $\bar{X}$ for that sample, and draws a yellow box directly above the value on the horizontal axis that is closest to the value of $\bar{X}$. The procedure is repeated until the tallest "stack" of boxes is too high to fit on the vertical axis. Run the lab again and watch how, one at a time, the boxes are drawn.

There are some important features of this graph. First, if you repeatedly run this lab, you will notice that the overall shape of the distribution of the means is bell-shaped. Second, the means are centered around the mean of the population ($\mu = 0$ in this case). Finally, the Empirical Rule tells us that because the distribution of these means is normal, approximately 95% of the means are between $\mu \pm 2\sigma/\sqrt{n} = 0 \pm 2\sigma/\sqrt{10} = \pm 0.63$.

Now change the value of n to 25 and run the lab. Overall, the shape of the distribution is the same. But now almost all of the means fall between $\mu \pm 2\sigma/\sqrt{25} = \pm 0.4$ (see Figure 7.5). This should not surprise you. Recall that the variance of the sampling distribution of $\bar{X}$ is σ^2/n. Therefore, as the sample size increases, the spread of the distribution will decrease.

We are now going to examine the shape of the sampling distribution of $\bar{X}$ for different parent distributions. Change the value of n back to 10, and select the Uniform(0,1) distribution under Sample from. For this distribution, the population mean is $\mu = 0.5$, and the population variance is $\sigma^2 = 1/12 = 0.0833$. Run the lab. The resulting graph should look something like the one shown in Figure 7.6. Notice that the sampling distribution is centered around the population mean $\mu = 0.5$.

Now, is $n = 10$ large enough to conclude that the distribution of the sample means is normal? The distribution does appear to be bell-shaped. If the distribution truly is normal, approximately 95% of the means will fall between $\mu \pm 2\sigma/\sqrt{10} = 0.5 \pm 2\sqrt{0.0833/10} = 0.5 \pm 0.18 = (0.32, 0.68)$. If we examine the graph, we see that this seems to be the case. Therefore, $n = 10$ is a large enough sample size to conclude that the distribution of the sample mean is normal.

FIGURE **7.5**
Distribution of means for samples with $n = 25$ from a Normal(0,1) distribution.

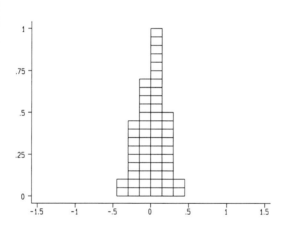

FIGURE **7.6**
Distribution of means for samples with $n = 10$ from a Uniform(0,1) distribution.

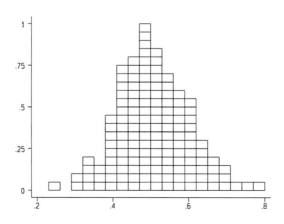

Is $n = 10$ large enough to make this statement when the parent distribution is an Exponential(1) distribution? Chose Exponential(1) under Sample from and run the lab for n = 10. By examining Figure 7.7, we see two things:

1 The distribution is centered at the population mean $\mu = 1$.
2 The shape of the distribution will not allow us to conclude that the sample means have a normal distribution because it is not bell-shaped.

Increase n to 50 and run the lab again. One possible result is shown in Figure 7.8. Again, the distribution is centered at the population mean $\mu = 1$. Also, for $n = 50$, the distribution appears to be bell-shaped. Let's use the Empirical Rule to see if normality can be justified. The interval containing 95% of the means is $\mu \pm 2\sigma/\sqrt{n} = 1 \pm 2(1/\sqrt{50}) = 1 \pm 0.28 = (0.72, 1.28)$. Based on Figure 7.8, this does not seem unreasonable. Therefore, $n = 50$ is large enough to conclude that the distribution of the sample mean is normal.

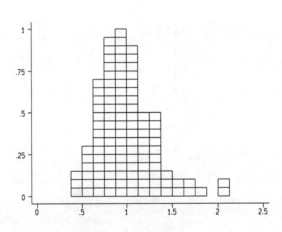

FIGURE 7.7
Distribution of means for samples with $n = 10$ from an Exponential(1) distribution.

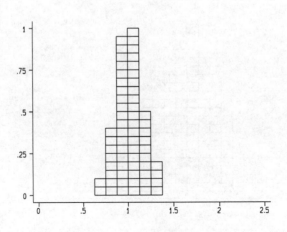

FIGURE 7.8
Distribution of means for samples with $n = 50$ from an Exponential(1) distribution.

7.4.3 500-Samples

Change Choice to 500-Samples, n to 10, and Sample from to Normal(0,1), and run the lab. For a brief time, you will see a bar that fills with blue as 500 samples are being selected and the sample mean for each sample is being computed. Then a yellow histogram of the distribution of the 500 means will be drawn. The corresponding normal distribution (with mean $\mu = 0$ and variance $\sigma^2/n = 1/10$) will be superimposed in red over the histogram. The resulting figure will look very similar to Figure 7.4. The histogram in your graph will differ from the one seen here because it is based on a different set of 500 samples. Notice how well the histogram fits the curve. The center of both is at $\mu = 0$, and the spread and shapes of both are very close. This should not surprise you. As we mentioned before, because the parent population is normally distributed, the sampling distribution of $\bar{X}$ is also normal. Therefore, we expect the histogram of the 500 means to closely resemble the theoretical normal curve.

Now let's examine how well the histogram fits the theoretical normal curve for samples of size $n = 10$ chosen from a Uniform(0,1) parent distribution. The normal curve will have mean $\mu = 0.5$ and variance $\sigma^2/n = 0.0833/10 = 0.00833$. Change Sample from to Uniform(0,1) and run the lab. The resulting figure will be similar to Figure 7.9. The histogram fits the theoretical curve well. This agrees with our observation in the previous section, that for samples from a Uniform(0,1) distribution, $n = 10$ is large enough to conclude that the sampling distribution of $\bar{X}$ is normal.

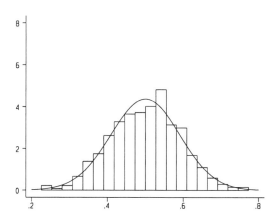

FIGURE 7.9
Normal curve superimposed on a histogram of 500 means of samples with $n = 10$ from a Uniform(0,1) population.

What happens when the distribution of the population is Exponential(1) and the sample size is n = 10? Earlier, we concluded that for an exponential(1) distribution, $n = 10$ was not large enough. Run the lab under these conditions. Our resulting graph is shown in Figure 7.10. The left side of the histogram is higher than the theoretical normal curve, and the right side is lower. This supports our earlier conclusion. The result of running this lab for n = 50 from an Exponential(1) distribution is shown in Figure 7.11. Not surprisingly, the histogram fits the theoretical normal curve.

FIGURE 7.10
Normal curve superimposed on a histogram of 500 means of samples with $n = 10$ from an Exponential(1) population.

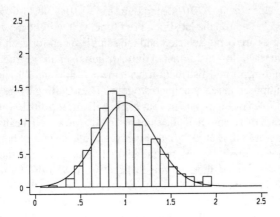

FIGURE 7.11
Normal curve superimposed on a histogram of 500 means of samples with $n = 50$ from an Exponential(1) population.

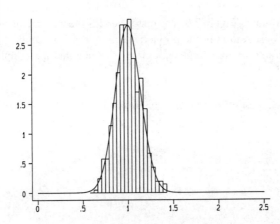

7.5 Summary

In this lab, we witnessed the Central Limit Theorem in action. Three properties of the sampling distribution of $\bar{X}$ were revealed:

1. The mean of the sampling distribution of $\bar{X}$ is the mean of the distribution of the population. The sample mean is *unbiased*, or "on target," as an estimator of the population mean. This behavior occurs regardless of the sample size.
2. The variance of the sampling distribution of $\bar{X}$ becomes smaller as the sample size increases. Thus, not only is the sample mean "on target," it hits closer to the target (μ) more *consistently* as the sample size increases. It becomes a more precise estimator of μ for larger values of n.
3. If n is "large enough," the sampling distribution of $\bar{X}$ is a normal distribution. How large the sample size needs to be depends on the distribution of the population

from which samples are drawn. The more skewed the parent population, the larger n needs to be to justify the normality statement.

7.6 Lab Exercises

7.6.1 Parent-Curves

1. Let Choice be Parent-Curves, run the lab, and print the graph. Briefly describe the shape of each distribution. What is the numerical value of the mean and the variance for each distribution? Which distributions are symmetrical and which are skewed?

7.6.2 One-at-a-time

For the exercises in this section, be sure that One-at-a-time is specified.

2. Let Sample from be Normal(0,1), and let n = 5. Run the lab repeatedly. Print the graph that you think best represents the overall patterns. What is the overall shape of the distribution of the means? Around what value is the distribution of the means centered? How is this value related to the mean of the population? Between what values do most of the means fall? How are these values related to the standard deviation of the population?

3. Let Sample from be Normal(0,1), and let n = 50. Run the lab repeatedly. Print the graph that you think best represents the overall patterns. What is the overall shape of the distribution of the means? Around what value is the distribution of the means centered? How is this value related to the mean of the population? Between what values do most of the means fall? How are these values related to the standard deviation of the population? Compare the shape and values to those in the previous exercise. Are there any similarities? What are the differences?

4. Repeat the previous exercise for n = 100. What happens to the shape of the distribution, the center of the distribution, and the spread of the distribution as the sample size increases?

5. Let Sample from be Uniform(0,1), and let n = 5. Run the lab, and print the resulting graph. What is the overall shape of the distribution of the sample means? What value is the distribution centered around? Use the Empirical Rule to determine the interval that approximately 95% of the means should fall between. Does this seem to be the case? Is n = 5 large enough to conclude that the sampling distribution of the sample means is normal when the parent population has a Uniform(0,1) distribution?

6. Let Sample from be Uniform(0,1). Starting with n = 5, run the lab six times, each time increasing the sample size by 1, until you reach n = 10. Print each of the resulting graphs. Using the Empirical Rule, what value of n is large enough to justify using the normal distribution for the sampling distribution of the mean when the parent population is Uniform(0,1)?

7. Let Sample from be Exponential(1). Starting with n = 10, run the lab, each time increasing the sample size by 5, until you reach a sample size that will allow you to conclude that the sampling distribution of the sample means is normal. Print each of the resulting graphs. (Hint: Use the Empirical Rule to determine what value of n is large enough to justify using the normal distribution for the sampling distribution of the mean when the parent population is Exponential(1)).

7.6.3 500-Samples

8. Let Choice be 500-Samples. Run the lab for each of the scenarios in the exercises in the previous section. Print all your graphs. What do you observe about the goodness of the fit of the histogram to the theoretical normal curve for each situation?

8

Z, t, χ^2, and F

In most introductory statistics courses, students study many different quantities, or *statistics*, calculated from a random sample. These statistics are used to calculate confidence intervals or to perform hypothesis tests to make inferences about some population parameter. A small set of fascinating results proves that under very general conditions, all of these statistics have a sampling distribution that belongs to one of four families of distributions.

8.1 Introduction

8.1.1 Some Basic Ideas

At first, it may seem odd to consider the sample mean as a random variable. But as we have seen in the *Random Sampling* and the *Central Limit Theorem* Labs, the value of the sample mean depends upon which elements are randomly chosen for the sample. Therefore, before the sample is selected, $\bar{X}$ is random. There are many possible samples of size n from a population and thus many possible values for $\bar{X}$. The distribution of all these values is called the *sampling distribution of* $\bar{X}$.

The same can be said for any statistic. Before a sample is selected from the population, we do not know what value we will calculate for the statistic. Because the sample is randomly selected, the statistic is a random variable. There are many samples of size n from the population, all of which will yield a slightly different value of the statistic. The function that describes the pattern of those values is the *sampling distribution* of the statistic. Remarkably, almost all of the statistics you will study in your introductory statistics courses have sampling distributions that belong to one of four families of distribution: normal, t, χ^2, and F. In the *How Are Populations Distributed?* Lab, you studied these four families. In this lab, our population is the set of all possible values of a statistic based on a random sample of size n. Here is a brief review of the four families.

8.1.2 The Normal Family

The normal family of distributions is the most important and best-known of all distributions. It is sometimes referred to as the "bell-shaped curve" because its graph looks like a bell. The parameter that determines the location of the center of the distribution on the real number line is the mean μ. The parameter that determines the height and width of the distribution is the variance σ^2. The *standard normal distribution* is a normal distribution with mean $\mu = 0$ and variance $\sigma^2 = 1$. If a random variable has a standard normal distribution, it is often denoted by Z.

8.1.3 The t Family

The t distribution is also a very important distribution. The sampling distribution of many statistics is a t distribution. A t distribution strongly resembles a standard normal distribution. It is centered at zero and bell-shaped. But the t distribution is not as tall as a standard normal distribution, and the tails of a t distribution are higher, or "thicker," than the tails of a normal distribution. The parameter that determines the height and width of a t distribution is called *degrees of freedom*, denoted by the lowercase Greek ν (nu). For a t distribution with ν degrees of freedom, we write t_ν. As the degrees of freedom for a t distribution get larger and larger, the t distribution gets closer and closer to the normal distribution.

8.1.4 The χ^2 Family

Although not as well known, the χ^2 (chi-square) distribution plays a fundamental role in statistics, too. The χ^2 distribution is skewed to the right. The degrees of freedom of a χ^2 distribution determine the center, height, and skewness of the distribution. Thus, for a χ^2 distribution with ν degrees of freedom, we write χ^2_ν.

8.1.5 The F Family

The F distribution is also a very well-known, extremely important distribution. The F distribution arises as the ratio of two χ^2 random variables. Therefore, the F distribution has two parameters, the *numerator degrees of freedom* ν_1 and the *denominator degrees of freedom* ν_2. As the names suggest, these are the degrees of freedom of the χ^2 random variables in the numerator and denominator of the ratio. These two parameters determine the shape of the F distribution. For an F distribution with numerator degrees of freedom ν_1 and denominator degrees of freedom ν_2, we write F_{ν_1, ν_2}.

8.1.6 Critical Values

Critical values are used in both confidence intervals and hypothesis testing. In confidence intervals, critical values are percentiles from the appropriate distribution that allow us to make the confidence statement. This will be explained in more detail in the *Interpreting Confidence Intervals* Lab (Chapter 11). In hypothesis testing, critical values are percentiles defining the rejection region that enable us to either reject or fail to reject the null hypothesis. The *Calculating Tests of Hypotheses* Lab (Chapter 14) will discuss this idea more thoroughly.

For now, we will describe the *level α critical value* as the number (or numbers) for which the total tail area(s) of the distribution determined by this number (or numbers) is equal to some specified value, α. Critical values are typically denoted by a letter (Z, t, χ^2, or F) representing the sampling distribution of the statistic. The letter has a numerical subscript indicating the area under the curve to the right of the critical value. For example, in a standard normal distribution, the value 1.96 "cuts off" the normal distribution, so that the area under the standard normal curve to the right of 1.96 is 0.025. Therefore, we would write $Z_{0.025} = 1.96$. This type of critical value is a *right-tailed level α critical value* because it cuts off only the right tail of the normal distribution.

In the same way, a *left-tailed level α critical value* cuts off the lower tail of the distribution, so to the right of the critical value, the area under the distribution is $1 - \alpha$. We can also have *two-tailed critical values*. For these, the total tail areas determined by the critical values sum to α. Consider $Z_{0.025} = 1.96$ and $Z_{0.975} = -1.96$. The tail area below -1.96 is 0.025, and the tail area above 1.96 is 0.025, so these two critical values together have level $\alpha = 0.05$. For critical values of distributions that have degrees of freedom, in the subscript, the tail area is given first, and the degrees of freedom are given second (for example, $t_{0.025, 10} = 2.28$).

8.2 Objectives

1. Visualize level α critical values for the sampling distributions Z, t, χ^2, and F.
2. Visualize a normal distribution and how that distribution is affected by changing values of μ and σ^2.
3. Visualize a t distribution and how that distribution is affected by changing values of the degrees of freedom.
4. Visualize a χ^2 distribution and how that distribution is affected by changing values of the degrees of freedom.
5. Visualize an F distribution and how that distribution is affected by changing values of the numerator and denominator degrees of freedom.
6. Understand the effects of n and π on the normal approximation to the binomial distribution.

8.3 Description

To begin this lab, select Z,t,Chi-square,F from the Labs menu. When you do, the following submenu will appear:

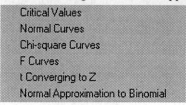

8.3.1 Critical Values

To open the *Critical Values* Lab, highlight and then click on Critical Values in the submenu. The dialog box shown in Figure 8.1 opens, giving a brief description of the lab. Below the description are five smaller boxes requiring information:

1. df_1: The value in this box specifies the degrees of freedom for the t or χ^2 distributions, or the numerator degrees of freedom for the F distribution. This may be ignored if the distribution specified in Distribution is Z. The initial value of df_1 is 20. The value of df_1 must be between 2 and 1000, and it must be an integer.

2. df_2: The value in this box specifies the denominator degrees of freedom for the F distribution. This may be ignored if the distribution specified in Distribution is Z, t, or Chi-square. The initial value of df_2 is 20. The value of df_2 must be between 2 and 1000, and it must be an integer.

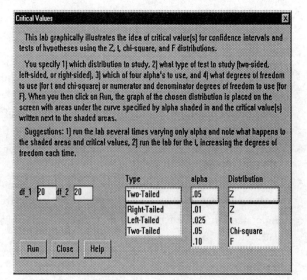

FIGURE **8.1**
Dialog box for the *Critical Values* Lab.

3   Type: There are three options for Type: Right-Tailed, Left-Tailed, and Two-Tailed. The choice that is specified will determine what type of critical value is displayed when the lab is run. The choices for Type are restricted to those given.

4   alpha: The value specified determines the tail area. It is the level of significance in a hypothesis test or 1 minus the level of confidence for a confidence interval. The initial value of alpha is 0.05. Because alpha is the area under a distribution, its values are restricted to 0 < alpha < 1. If you want a value of alpha that is not given as one of the options, simply type the desired value of alpha in the box.

5   Distribution: The options given below Distribution determine which distribution—Z, t, χ^2, or F—is to be studied. The choices for Distribution are restricted to those given. The initial distribution is the Z, or standard normal, distribution.

Below df_1 and df_2 are three buttons:

1   Run: Clicking on this button causes the lab to calculate critical values for the specified values in df_1, df_2, Type, alpha, and Distribution. A graphics window opens containing a graph of the distribution drawn in red and the critical values with the tail areas shaded in yellow. An example of this graph for the initial values is shown in Figure 8.2. This graph will be discussed further in Section 8.4.

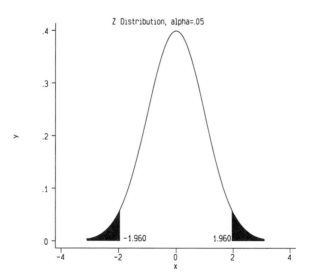

FIGURE 8.2
Two-tailed critical values from the Z distribution for alpha = 0.05.

2   Close: Clicking on this button closes the lab and returns you to the StatConcepts menus.

3   Help: Clicking on this button opens a help window containing information about the lab.

8.3.2 Normal Curves

To start this lab, highlight and click on Normal Curves in the submenu under Z, t, Chi-square, F. On the left of the screen is the small dialog box shown in Figure 8.3. To the right of the dialog box is a graphics window containing a graph of the standard normal distribution drawn in yellow (see Figure 8.4). This dialog box contains seven buttons:

1. Reset: Clicking on this button erases any previously drawn distributions and returns the value of the mean to 0 and the variance to 1. The standard normal distribution is redrawn in yellow in the graphics window.

2. mu +: Clicking on this button increments the current value of the mean by 1. The value of Mean is updated to reflect the current value of the mean. A graph of the normal distribution with the new value of the mean is drawn in yellow. The previous normal distribution is redrawn in red.

FIGURE 8.3
Dialog box for the Normal Curves Lab.

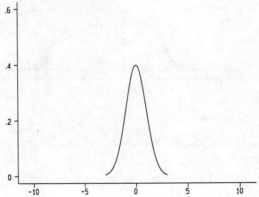

FIGURE 8.4
Standard normal distribution ($\mu = 0, \sigma^2 = 1$).

3. **mu -**: Clicking on this button decreases the current value of the mean by 1. The value of Mean is updated to reflect the current value of the mean. A graph of the normal distribution with the new value of the mean is drawn in yellow. The previous normal distribution is redrawn in red.

4. **var +**: Clicking on this button increments the current value of the reciprocal of the standard deviation (remember, $\sigma = \sqrt{\sigma^2}$) by 0.1. The value of Variance is updated to reflect the current value of the variance. A graph of the normal distribution with the new value of the variance is drawn in yellow. The previous normal distributions are redrawn in red.

5. **var -**: Clicking on this button decreases the current value of the reciprocal of the standard deviation (remember, $\sigma = \sqrt{\sigma^2}$) by 0.1. The value of Variance is updated to reflect the current value of the variance. A graph of the normal distribution with the new value of the variance is drawn in yellow. The previous normal distributions are redrawn in red.

6. **Help**: Clicking on this button opens a help window containing information about the lab.

7. **Close**: Clicking on this button closes the lab and returns you to the StatConcepts menus.

Below these buttons, the current values of Mean and Variance are given. Because the initial distribution is a standard normal distribution, the initial values of Mean and Variance are 0 and 1, respectively.

8.3.3 Chi-square Curves

To start this lab, highlight and click on Chi-square Curves in the submenu under Z, t, Chi-square, F. On the left of the screen is the small dialog box shown in Figure 8.5. To the right of the dialog box is a graphics window containing a graph of the χ^2_{10} distribution drawn in yellow (see Figure 8.6). This dialog box contains five buttons:

1. **Reset**: Clicking on Reset erases any previously drawn distributions and returns the value of degrees of freedom back to 10. The χ^2_{10} is redrawn in yellow in the graphics window.

2. **df +**: Clicking on this button increments the current value of the degrees of freedom by the value in Inc. The value of df is updated to reflect the current value of the degrees of freedom. A graph of the χ^2 distribution with the new value of degrees of freedom is drawn in yellow. The previous χ^2 distribution is redrawn in red. The maximum value allowed for total degrees of freedom is 100. If you increment degrees of freedom to be larger than 100, a graph of the χ^2 distribution will be drawn as if the degrees of freedom were 100.

3. **df -**: Clicking on this button decreases the current value of the degrees of freedom by the value in Inc. The value of df is updated to reflect the current value of the degrees of freedom. A graph of the χ^2 distribution with the new

FIGURE 8.5
Dialog box for the *Chi-square Curves* Lab.

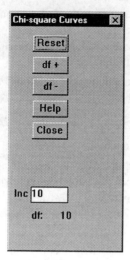

FIGURE 8.6
χ^2_{10} distribution.

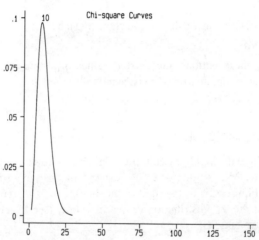

value of degrees of freedom is drawn in yellow. The previous χ^2 distribution is redrawn in red. The minimum value allowed for total degrees of freedom is 10. If you increment degrees of freedom to be smaller than 10, a graph of the χ^2 distribution will be drawn as if the degrees of freedom were 10.

4 Help: Clicking on this button opens a help window containing information about the lab.

5 Close: Clicking on this button closes the lab and returns you to the StatConcepts menus.

Below these buttons is a small box labeled Inc. The value in this box specifies the change in degrees of freedom when df + or df − is clicked. The initial value for Inc is 10, and the value of Inc must be an integer. For this lab, degrees of freedom cannot be smaller than 10 or larger than 100. Below Inc is the current value of degrees of freedom. The initial value for degrees of freedom is df = 10.

8.3.4 *F* Curves

To start this lab, highlight and click on F Curves in the submenu under Z, t, Chi-square, F. On the left of the screen is the small dialog box shown in Figure 8.7. To the right of the dialog box is a graphics window containing a graph of the $F_{10,10}$ distribution drawn in yellow (see Figure 8.8). This dialog box contains seven buttons:

1. Reset: Clicking on this button erases any previously drawn distributions and returns the value of both numerator and denominator degrees of freedom to 10. The $F_{10,10}$ distribution is redrawn in yellow in the graphics window.

2. df1 +: Clicking on this button increments the current value of the numerator degrees of freedom by the value in Inc. The value of df_1 is updated to reflect the current value of the numerator degrees of freedom. A picture of the *F* distribution with the new value of numerator degrees of freedom is drawn in yellow.

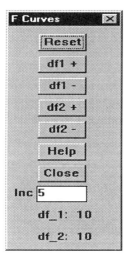

FIGURE **8.7**
Dialog box for the *F Curves* Lab.

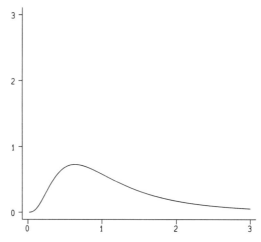

FIGURE **8.8**
$F_{10,10}$ distribution.

The previous F distribution is redrawn in red. The maximum value allowed for numerator degrees of freedom is 200. If you increment numerator degrees of freedom to be larger than 200, a graph of the F distribution will be drawn as if the numerator degrees of freedom were 200.

3. `df1 -`: Clicking on this button decreases the current value of the numerator degrees of freedom by the value in `Inc`. The value of `df_1` is updated to reflect the current value of the numerator degrees of freedom. A graph of the F distribution with the new value of numerator degrees of freedom is drawn in yellow. The previous F distribution is redrawn in red. The minimum value allowed for numerator degrees of freedom is 10. If you increment numerator degrees of freedom to be smaller than 10, a graph of the F distribution will be drawn as if the numerator degrees of freedom were 10.

4. `df2 +`: Clicking on this button increments the current value of the denominator degrees of freedom by the value in `Inc`. The value of `df_2` is updated to reflect the current value of the denominator degrees of freedom. A graph of the F distribution with the new value of denominator degrees of freedom is drawn in yellow. The previous F distribution is redrawn in red. The maximum value allowed for denominator degrees of freedom is 200. If you increment denominator degrees of freedom to be larger than 200, a graph of the F distribution will be drawn as if the denominator degrees of freedom were 200.

5. `df2 -`: Clicking on this button decreases the current value of the denominator degrees of freedom by the value in `Inc`. The value of `df_2` is updated to reflect the current value of the denominator degrees of freedom. A graph of the F distribution with the new value of denominator degrees of freedom is drawn in yellow. The previous F distribution is redrawn in red. The minimum value allowed for denominator degrees of freedom is 10. If you increment denominator degrees of freedom to be smaller than 10, a graph of the F distribution will be drawn as if the denominator degrees of freedom were 10.

6. `Help`: Clicking on this button opens a help window containing information about the lab.

7. `Close`: Clicking on this button closes the lab and returns you to the StatConcepts menus.

Below these buttons is a small box labeled `Inc`. The value in this box specifies the change in degrees of freedom when `df1 +`, `df1 -`, `df2 +`, or `df2 -` is clicked. The initial value for `Inc` is 5, and the value of `Inc` must be an integer. For this lab, degrees of freedom cannot be smaller than 10 or larger than 200. Below `Inc` is the current value of both degrees of freedom. The initial value for degrees of freedom is 10.

8.3.5 t Converging to Z

To start this lab, highlight and click on `t Converging to Z` in the submenu under `Z, t, Chi-square, F`. On the left of the screen is the small dialog box shown in Figure 8.9, which contains three boxes requiring input:

FIGURE **8.9**
Dialog box for the
t Converging to Z Lab.

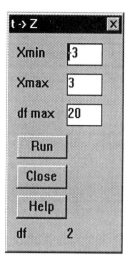

1 Xmin: The value in the box to the right of Xmin determines the lower value of the horizontal axis of the graph. Graphs of the *t* and *Z* distribution will be displayed only for values of a random variable that are greater than Xmin. The initial value of Xmin is −3.

2 Xmax: The value in the box to the right of Xmax determines the upper value of the horizontal axis on the graph. Graphs of the *t* and *Z* distribution will be displayed only for the values of a random variable that are less than Xmax. The initial value of Xmax is 3.

3 df max: The value in the box to the right of df max determines the maximum degrees of freedom for the final *t* distribution. The initial value of df max is 20. Therefore, the final distribution that will be displayed for this initial value is a t_{20} distribution.

Below these input areas are three buttons:

1 Run: Clicking on this button opens a graphics window. The contents of this window are described further in Subsection 8.4.5.

2 Close: Clicking on this button closes the lab and returns you to the StatConcepts menus.

3 Help: Clicking on this button opens a help window containing information about the lab.

Beneath these three buttons, df appears, with a number to the right indicating the degrees of freedom of the *t* distribution currently being displayed in the graphics window that opens when Run is clicked.

8.3.6 Normal Approximation to Binomial

Choosing this concept lab opens the dialog box shown in Figure 8.10, which briefly discusses when it is appropriate to use the normal distribution to approximate binomial probabilities. In the bottom left-hand corner are two boxes where you can enter desired

FIGURE 8.10
Dialog box for the *Normal Approximation to Binomial Probabilities* Lab.

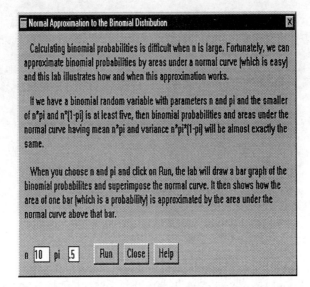

values for n and pi. For this lab, the values of n must be between 2 and 49. Initial values are n = 10 and pi = 0.50. To the right of the boxes are three buttons:

1. Run: Clicking on this button opens a graphics window containing the binomial probability distribution for the specified values of n and pi, with the normal curve approximation superimposed (see Figure 8.11). The figure also shows how the area of one bar from the binomial distribution is approximated by the area under the normal curve above that bar.

2. Close: Clicking on this button closes the lab and returns you to the StatConcepts menus.

3. Help: Clicking on this button opens a help window containing information about the lab.

FIGURE 8.11
Example of the *Normal Approximation to Binomial* Lab.

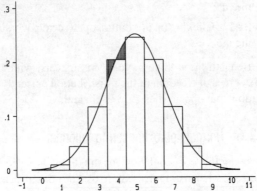

8.4 Guided Tour of the Lab

The tour of this lab will provide you with an abundance of new information and ideas. We will make several stops, each of which has its own unique sights.

8.4.1 Critical Values

Begin this lab by clicking on `Critical Values` from the submenu `Z,t, Chi-square,F`. Running the lab for the initial values will result in the graph shown in Figure 8.2. The curve in red is a picture of the Z, or standard normal, distribution. The areas shaded in yellow are the tail areas that are "cut off" by the critical values 1.96 and −1.96. The sum of these areas is `alpha = 0.05`. Because `Type` is `Two-Tailed`, both the upper and lower tails of the distribution are used. Two-tailed critical values are used for confidence intervals or in the rejection region of hypothesis tests for which the alternative hypothesis has a "$\neq$" sign.

We want to choose critical values for two-tailed procedures so that the upper and lower tail areas are equal. Because `alpha` must total 0.05, this means that the areas are both 0.025. Because any normal distribution is symmetrical about its mean, these areas are mirror images of each other (we will see this is not the case for all distributions). And because the mean of the standard normal distribution is 0, the upper and lower critical values are equal in magnitude (both are 1.96), but the upper critical value is positive and the lower critical value is negative. Thus, we write $Z_{0.975} = -1.96$ and $Z_{0.025} = 1.96$.

Now change `Type` to `Right-Tailed`. Right-tailed procedures are most commonly used in the rejection regions for hypothesis tests with an alternative containing a ">" sign. We still want the tail area to be `alpha = 0.05`, but now we are only interested in the right tail. In Figure 8.12, notice that the shaded area is now only in the right tail of the standard normal distribution. The total tail area is still 0.05, but now it is all on the same side of the distribution. Therefore, $Z_{0.05} = 1.645$; that is, 1.645 is the value on the real number line, so the area under the standard normal distribution to the right of 1.645 is 0.05. If you change `Type` to `Left-Tailed` and run the lab, the result will be a mirror image of what we just saw, except that the critical value

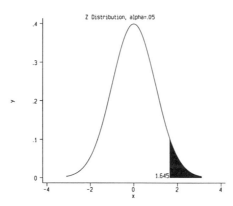

FIGURE **8.12**
Right-tailed critical value from the Z distribution for `alpha = 0.05`.

will be $Z_{0.95} = -1.645$ (see Figure 8.13). Rejection regions in hypothesis tests with a "<" sign in the alternative hypothesis are the most common use for lower-tailed critical values.

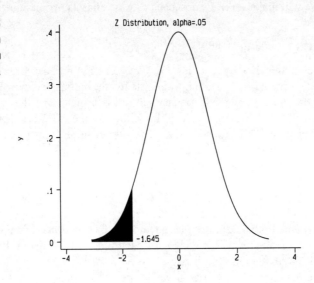

FIGURE 8.13
Left-tailed critical value from the Z distribution for alpha = 0.05.

Now let's begin to examine the t distribution and how changing the degrees of freedom affects the critical values of that distribution. In Subsection 8.4.5, we will see even more about the behavior of the t distribution for increasing degrees of freedom. But for now, let df_1 = 5, Type be Two-Sided, alpha = 0.05, and Distribution be the t distribution, and run the lab. Notice in Figure 8.14 that $t_{0.975,5} = -2.571$ and $t_{0.025,5} = 2.571$. Both tail areas are 0.025, for a total tail area

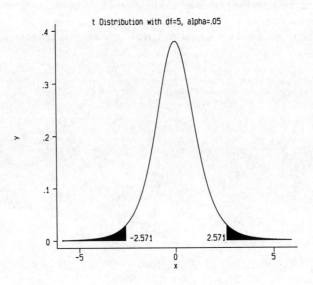

FIGURE 8.14
Two-tailed critical value from the t_5 distribution for alpha = 0.05.

of 0.05. Like the standard normal distribution, the t distribution is symmetrical about zero, so the two areas appear to be mirror images of each other.

Now run the lab for the same Type, alpha, and Distribution, but let df_1 = 20. We see in Figure 8.15 that $t_{0.975, 20} = -2.086$ and $t_{0.025, 20} = 2.086$. Next, increase df_1 to 50. In Figure 8.16, the critical values have become $t_{0.975, 50} = -2.009$ and $t_{0.025, 50} = 2.009$. The critical values for the t distribution keep getting closer and closer to the two-sided level $\alpha = 0.05$ critical values for the Z distribution!

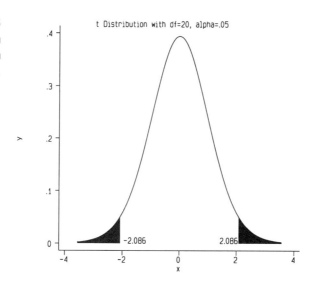

FIGURE 8.15
Two-tailed critical value from the t_{20} distribution for alpha = 0.05.

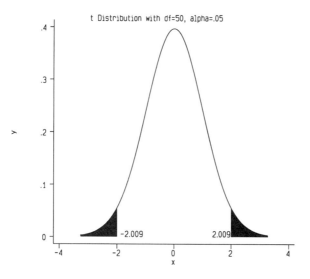

FIGURE 8.16
Two-tailed critical value from the t_{50} distribution for alpha = 0.05.

How do critical values for the χ^2 distribution behave? The χ^2 distribution is not symmetrical, but rather is skewed. Further, the χ^2 distribution begins at zero.

Let `df_1` = 20, `alpha` = 0.05, `Type` be `Two-Sided`, and `Distribution` be `Chi-square`. The resulting graph is shown in Figure 8.17. The two critical values are $\chi^2_{0.975,20} = 9.591$ and $\chi^2_{0.025,20} = 34.170$. Although both tail areas equal 0.025, they are *not* mirror images of each other. Further, the lower critical value is not the negative of the upper critical value. This behavior is typical for skewed sampling distributions.

The F distribution also begins at zero and is positively skewed. Run the lab for `Distribution F` with `df_1` and `df_2` = 20, `alpha` = 0.05, and `Type Two-Tailed`. You can see in Figure 8.18 that the behavior of F critical values is similar to that of χ^2 critical values. In this graph, the lower critical value is $F_{0.975,20,20} = 0.406$ and the upper critical value is $F_{0.975,20,20} = 2.464$. Again, both tail areas are 0.025, but they are not mirror images of each other.

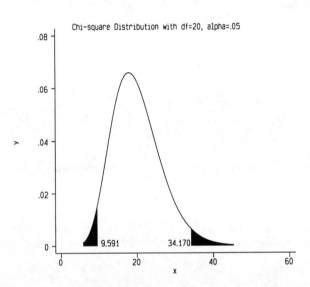

FIGURE 8.17
Two-tailed critical value from the χ^2_{20} distribution for `alpha` = 0.05.

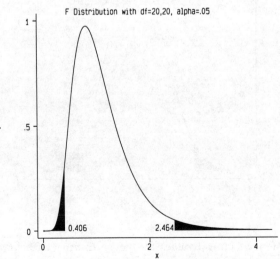

FIGURE 8.18
Two-tailed critical value from the $F_{20,20}$ distribution for `alpha` = 0.05.

8.4.2 Normal Curves

The normal distribution, or "bell-shaped" curve, is the most important, most widely used, and best-known distribution in statistics. Two parameters, the mean μ and the variance σ^2, completely determine the center, height, and width of the distribution. In this lab, we will study exactly how the normal distribution is affected by changing values of μ and σ^2.

The mean μ of the normal distribution is the exact center of the distribution. From the submenu Z, t, Chi-square, F, highlight and click on Normal Curves, and run the lab. The resulting graph is shown in Figure 8.4. The curve in red is a graph of the *standard normal distribution*. This particular normal distribution has mean $\mu = 0$ and variance $\sigma^2 = 1$. Notice that the center of the distribution is at $\mu = 0$. Now, to see what happens when the mean changes, click on mu + (see the dialog box in Figure 8.3). This increases the value of μ by 1 and gives a graph of the new normal distribution (with mean $\mu = 1$) in yellow. The red distribution is the standard normal distribution (with mean $\mu = 0$). Notice in Figure 8.19 that the height and width are exactly the same for the two distributions. Changing the mean simply shifts the entire distribution so that the center is at the new value of the mean, in this case, 1. Click on mu − twice. The mean is now $\mu = -1$. The entire distribution has shifted to the left two units and is now centered at $\mu = -1$, but the height and width remain unaffected (see Figure 8.20).

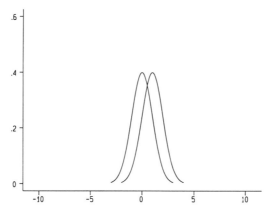

FIGURE **8.19** Comparison of the standard normal distribution and the normal distribution with mean $\mu = 1$ and variance $\sigma^2 = 1$.

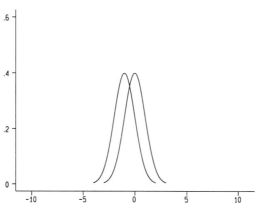

FIGURE **8.20** Comparison of the standard normal distribution and the normal distribution with mean $\mu = -1$ and variance $\sigma^2 = 1$.

Now click on Reset. This will erase everything so that we can examine the effect of changing the variance on the distribution. Clicking on var + increases the variance, causing the distribution to become shorter and wider, but the location of the center of the distribution does not change (see Figure 8.21). Conversely, clicking on var - twice decreases the variance, causing the distribution to become taller and thinner (see Figure 8.22). The reason for these changes is that the total area of the distribution must be 1. The variance measures the spread of the distribution. Therefore, if the distribution is getting more spread out (the variance is increasing), the height must be getting smaller so that the total area will not change.

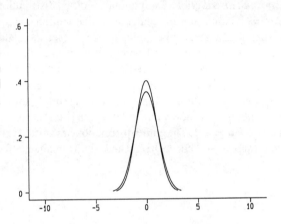

FIGURE 8.21 Comparison of the standard normal distribution and the normal distribution with mean $\mu = 0$ and variance $\sigma^2 = 1.23$.

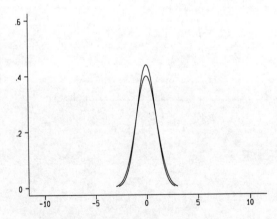

FIGURE 8.22 Comparison of the standard normal distribution and the normal distribution with mean $\mu = 0$ and variance $\sigma^2 = 0.83$.

8.4.3 Chi-square Curves

Start this lab by clicking on Chi-square Curves in the Z, t, Chi-square, F submenu. The graph you see drawn in yellow is the χ^2_{10} distribution, the chi-square distribution with 10 degrees of freedom. This graph is shown in Figure 8.6. Notice that the distribution is centered at $\mu = 10$. Leave Inc = 10 and click on df + in the dialog box. The degrees of freedom are incremented by 10, and the χ^2_{20} distribution is drawn in yellow and the χ^2_{10} distribution is redrawn in red (see Figure 8.23). There are a few things to notice about this graph:

1. The center of the distribution has shifted to $\mu = 20$. It turns out that the degrees of freedom for χ^2 are the mean of the χ^2 distribution.

2. The distribution is wider and shorter for 20 degrees of freedom than for 10. This is because the variance of a χ^2 distribution is $\sigma^2 = 2\nu = 2(20) = 40$. Thus, as degrees of freedom increase, the variance of the distribution doubles. Because the area under the curve must equal 1, if the distribution is getting wider, it must also get shorter so that the area is not affected.

3. The distribution is more symmetrical for larger degrees of freedom.

Click on df + four more times until you get the same graph as in Figure 8.24. The distribution is less skewed, it is wider and shorter (the variance is $\sigma^2 = 120$), and its mean is at its degrees of freedom ($\mu = 60$).

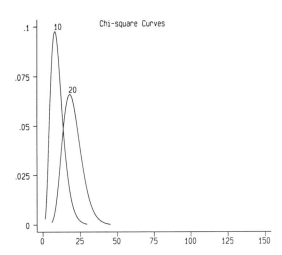

FIGURE 8.23 χ^2_{20} distribution.

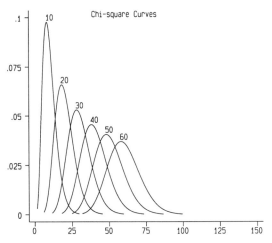

FIGURE 8.24 χ^2_{60} distribution.

8.4.4 F Curves

In terms of the effect that changing degrees of freedom has on the shape of the distribution, the F distribution is the most complex distribution we will study in this tour. We will examine the effect of changing the numerator degrees of freedom first, then the denominator degrees of freedom, then both.

To start the lab, highlight and click on F Curves from the submenu Z, t, Chi-square, F. The distribution shown in Figure 8.8 is an $F_{10,10}$ distribution. For reference, the value of the mean of this distribution is $\mu = 1.25$, and the variance is $\sigma^2 = 0.9375$. Let's examine how the location and shape of this distribution change as the numerator degrees of freedom change. Click on df1 + in the dialog box. The numerator degrees of freedom will be incremented by the value of Inc = 5; the denominator degrees of freedom remain unaffected. A graph of the new F distribution is drawn in yellow, and the $F_{10,10}$ distribution is redrawn in red (see Figure 8.25). Click on df1 + again. The new distribution is an $F_{20,10}$ distribution (see Figure 8.26). Let's examine closely these three distributions.

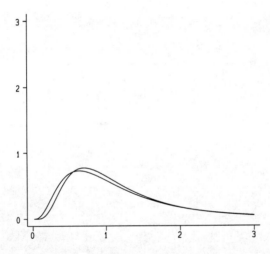

FIGURE 8.25 $F_{10,10}$ and $F_{15,10}$ distributions.

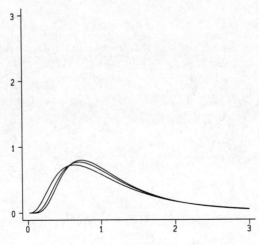

FIGURE 8.26 $F_{10,10}$, $F_{15,10}$, and $F_{20,10}$ distributions.

Recall from previous labs that the mean of a distribution is the location of the center of mass of the distribution, or its balance point. Does it appear in Figure 8.26 that the center of mass has shifted for these three distributions as the numerator degrees of freedom have increased? In fact, it has not. For all three distributions, the value of the mean is $\mu = 1.25$! It turns out that the expression for calculating the mean of an F_{ν_1,ν_2} distribution is

$$\mu = \frac{\nu_2}{\nu_2 - 2}. \tag{8.1}$$

This does not even involve ν_1, so the mean is unaffected by changes in ν_1.

Also notice that the preceding formula places a constraint on the values allowable for ν_2. First, the mean of an F distribution cannot be negative, because the F distribution itself is strictly positive. Second, we cannot divide by zero. Therefore, $\nu_2 > 2$.

The spread of any distribution is measured by the variance, so what about the variance? Has it changed as ν_1 has increased? In fact, it has. For the $F_{15,10}$ distribution, the variance is $\sigma^2 = 0.7986$, and for the $F_{20,10}$ distribution, it is $\sigma^2 = 0.7292$. The expression for the variance of an F_{ν_1,ν_2} distribution is a somewhat complicated function of ν_1 and ν_2:

$$\sigma^2 = \frac{2\nu_2^2(\nu_1 + \nu_2 - 2)}{\nu_1(\nu_2 - 2)^2(\nu_2 - 4)}. \tag{8.2}$$

From this, we get a second, more strict constraint on the value of ν_2: It must be that $\nu_2 > 4$.

Now click on Reset. We are going to see what happens to the F distribution when the numerator degrees of freedom remain unchanged and the denominator degrees of freedom change. These changes will be both more complex and more subtle. Change the value of Inc from 5 to 10 and click on df2 + one time. In the resulting graph, shown in Figure 8.27, does it appear that the center of mass, the mean, has changed? It should appear to have decreased. The mean for the $F_{10,20}$ distribution is $\mu = 1.1111$.

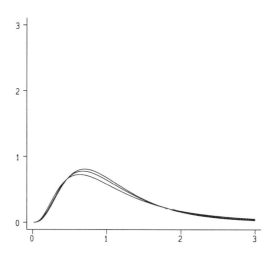

FIGURE **8.27**
$F_{10,10}$ and $F_{10,20}$ distributions.

A second examination of the formula for the mean of an F distribution explains some of the features we have been observing.

1. Because ν_1 is not a part of this expression, changes in ν_1 will not affect the mean.
2. Because $\nu_2 - 2$ is in the denominator, and the F distribution is strictly positive, $\nu_2 > 2$. Therefore, if we restrict ν_2 to the set of integers, the largest the mean of an F distribution can be is $\mu = 3/(3-2) = 3$.
3. Because the mean is a ratio of ν_2 and $\nu_2 - 2$, as ν_2 gets larger, the mean will decrease. Further, the larger ν_2 gets, the closer μ gets to 1.

Now compare the widths and heights of the two distributions. Again using the formula, we can calculate $\sigma^2 = 0.4321$ for the $F_{10,20}$ distribution. Simply looking at this expression, it may not be immediately clear how the variance behaves as the denominator degrees of freedom increase. It turns out that if ν_2 gets very large, the variance of the F distribution gets closer and closer to $2/\nu_1$.

Once again, click on Reset. This time, let Inc = 10. Beginning with df1 +, alternate clicking on df1 + and df2 + until you have clicked on both buttons three times. The resulting graph is shown in Figure 8.28. Changes in ν_2 have a slightly larger effect on the shape of the F distribution than changes in ν_1, but from what we have seen, this is not surprising. The overall effect of the changes is that the center of the distribution is moving gradually toward 1 while the distribution itself is becoming taller and thinner.

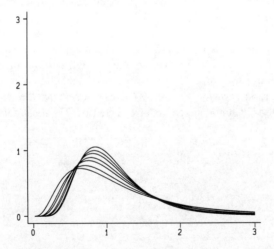

FIGURE 8.28 Comparison of seven F distributions.

8.4.5 t Converging to Z

The t distribution is remarkably similar to the Z distribution. Select t Converging to Z from the Z, t, Chi-square, F submenu, and change df max to 5 in the dialog box. Then run the lab. The first curve that appears in red is a graph of the standard normal distribution. Beginning with 1 degree of freedom, a graph of the t_1 distribution is drawn in blue. Then degrees of freedom are incremented by 1, the previous t distribution is erased, and the new t_2 distribution is drawn in blue. This process continues until we reach df max = 5. The resulting final graph is shown in Figure 8.29. There are several features of this graph you should notice. First, both distributions have the same overall bell shape, and both are centered at zero. However, the tails of the t distribution (drawn in blue) are higher, or thicker, than the tails of the standard normal distribution. Because both distributions must have a total area of 1, this means that the t distribution will have to be shorter than the standard normal distribution.

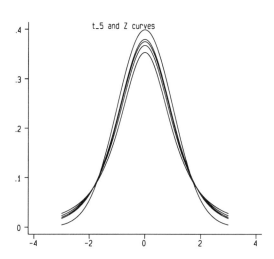

FIGURE **8.29**
t_5 distribution compared to a standard normal distribution.

We can examine the difference in the heights of the distributions more closely by restricting Xmin to −1 and Xmax to 1. Make these changes and run the lab. The result is shown in Figure 8.30. Notice how much shorter the blue t_5 distribution is than the red Z distribution. We can also examine the difference in the heights of the upper tails by letting Xmin = 2 and Xmax = 4 (see Figure 8.31). Notice how much taller the blue tails of the t_5 distribution are than the red tails of the Z distribution.

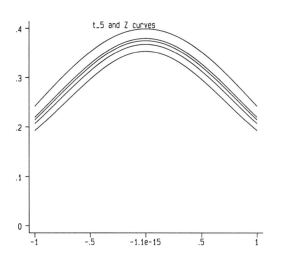

FIGURE **8.30**
Comparison of the centers of a t_5 and a standard normal distribution.

Now run the lab for df max = 20. Degrees of freedom will be incremented by 1 and successive t distributions will be drawn until we reach 20 degrees of freedom. While the graphs are being drawn, notice how as the degrees of freedom for the t distribution increase, the t distribution gets taller and its tails become less thick. Figure 8.32 shows the final graph. In fact, if we could continually draw t distributions until we reached an infinite number of degrees of freedom, we would see that the t distribution would eventually be exactly the standard normal distribution!

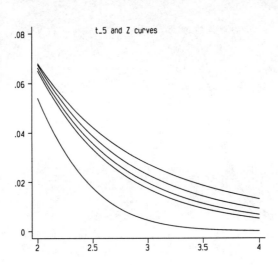

FIGURE **8.31**
Comparison of the upper tails of a t_5 and a standard normal distribution.

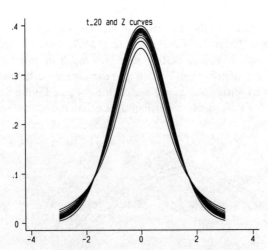

FIGURE **8.32**
t_{20} distribution compared to a standard normal distribution.

8.4.6 Normal Approximation to Binomial

Until now, our tour has dealt with the sampling distribution of a statistic that was calculated from a sample drawn from a continuous population. Now we will see what happens when the parent population is a 0–1 population. We have previously visited this lab in the *Sampling From 0–1 Populations* Lab (Chapter 5). This time, we examine it from a different perspective.

When a random sample is selected from a 0–1 population, the parameter of interest is the population proportion π of 1's. The statistic used to estimate π is

$$p = \frac{\text{\# of 1's}}{n}.$$

Because the only values we can observe from a 0–1 population are 0 and 1, p can also be calculated using

$$p = \frac{1}{n} \sum_{i=1}^{n} x_i.$$

But this is exactly the sample mean. Thus, we can use a modified version of the Central Limit Theorem (the *Central Limit Theorem* Lab examines this result thoroughly).

When we are sampling from a 0–1 population, if n is too small, the sampling distribution of np is a binomial distribution. However, if the smaller of the expected number of 1's ($n\pi$) and the expected number of 0's ($n(1-\pi)$) is at least 5, then we can use the normal distribution to approximate the binomial distribution! This allows us to use the normal distribution for purposes of statistical inference on π. Because the binomial distribution is discrete and the normal distribution is continuous, using this approximation, when it is appropriate, actually allows more flexibility in which values of α we can choose. For example, if $n = 20$ and $\pi = 0.5$, there is no critical value from the binomial distribution such that *exactly* 0.05 of the area falls above that value. The closest we can get to 0.05 is a critical value with a tail area of 0.0882. Therefore, the resulting confidence interval would not be the 95% confidence interval, but rather an 82.36% confidence interval. So when will this approximation yield reliable inferences, and when will it not?

Run the lab by choosing Normal Approximation to Binomial from the Z, t, Chi-square, F submenu. You will see a dialog box (Figure 8.10) explaining how to use this lab, and then a graph like the one shown in Figure 8.11. This is a graph of the binomial distribution for $n = 10$ and $\pi = 0.50$ with a graph of the normal distribution with mean $\mu = n\pi = 5$ and variance $\sigma^2 = n\pi(1-\pi) = 2.5$ superimposed. In the *Sampling From 0–1 Populations* Lab, we studied this graph in detail. To refresh your memory, we will briefly recap that discussion. If we use the binomial probability distribution function to calculate $P(X = 4)$, the result is $P(X = 4) = 0.2051$. However, for the normal distribution, the result is $P(X = 4) = 0$, because it is continuous. Thus, to use the normal distribution to approximate this discrete probability, we calculate $P(3.5 < X < 4.5) = 0.2045$. This approximation works well because even though we "chop off" the area of the bar for the binomial distribution from 3.5 to 4, we "add in" almost the same amount of area under the curve from 4 to 4.5. The difference in the exact binomial probability (0.2051) and the approximated probability (0.2045) is equal to the difference in the "chopped off" area and the "added in" area. This difference is only 0.0006!

If we used the normal approximation for critical values, there would be an additive effect. That is, the difference 0.0006 applies only to one bar in the binomial distribution. There will be a slight difference between the normal approximation and the binomial distribution for each bar involved in the area we are approximating. This could drastically affect the reliability of our inference if the assumptions are not satisfied.

Run the lab for pi = 0.20 and n = 10. The smaller of $n\pi$ and $n(1-\pi)$ is only 2. The resulting graph, shown in Figure 8.33, tells the story well. It is very easy to see that the "chopped off" area is much, much larger than the "added in" area. The difference in the two areas is $0.2285 - 0.2684 = -0.0399$. Imagine this effect accumulated over two or three bars! When you consider this, it is easy to see the importance of ensuring that the smaller of $n\pi$ and $n(1-\pi)$ is at least 5. If this assumption is not satisfied, the results of an inference procedure based on the normal approximation could be terribly misleading.

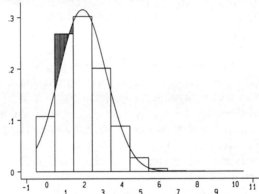

FIGURE 8.33
Normal approximation to binomial probabilities for n = 10 and pi = 0.20.

8.5 Summary

There were a number of sights to see on this tour, including these main attractions:

1. Critical values are numbers that "cut off" the tail of a distribution so that the area in that tail is equal to a specified value α. Critical values have three types:

    (a) *Right-tailed:* The right tail area specified by the critical value is equal to α. Right-tailed critical values are most often used in rejection regions for hypothesis tests that have an alternative hypothesis with a ">" sign.

    (b) *Left-tailed:* The left tail area specified by the critical value is equal to α. Left-tailed critical values are most often used in rejection regions for hypothesis tests that have an alternative hypothesis with a "<" sign.

    (c) *Two-tailed:* Two-tailed critical values determine left and right tails with area $\alpha/2$ so that the total tail area is α. Two-tailed critical values are used for confidence intervals and in rejection regions for hypothesis tests having a "$\neq$" sign in the alternative. If the distributions are symmetrical about zero (like the standard normal and t distributions), the left critical value will be equal to the negative of the right critical value and the two tail areas will be mirror images of each other.

2. The normal distribution has two parameters that completely determine the shape of the distribution. The mean affects only the location of the center. Increasing the mean shifts the distribution to the right; decreasing the mean shifts the distribution to the left. The variance affects the height and width of the distribution: the larger the variance, the flatter and more spread out the distribution.

3. The χ^2 distribution begins at zero and is skewed to the right. The larger the degrees of freedom become, the more symmetrical the distribution appears. The mean of a χ_ν^2 distribution is ν.

4. The F distribution begins at zero and is skewed to the right. It is a very complex distribution characterized by several features:

    (a) If the denominator degrees of freedom remain unchanged, the mean of the F distribution remains unchanged; the mean of an F distribution is unaffected by changes in numerator degrees of freedom. The denominator degrees of freedom completely determine the mean. The largest value for the mean occurs at $\nu_2 = 3$ and is $\mu = \nu_2/(\nu_2 - 2) = 3/(3 - 2) = 3$. The larger ν_2 becomes, the smaller the mean becomes. However, as the increase in ν_2 becomes larger, the decrease in μ becomes smaller. If ν_2 were very large, the mean of the F distribution would be very close to 1.

    (b) The variance of an F_{ν_1,ν_2} distribution behaves in a somewhat complicated manner. However, as the denominator degrees of freedom get very large, the variance of the F_{ν_1,ν_2} distribution gets very close to $2/\nu_1$.

    (c) When both numerator and denominator degrees of freedom are changing, the change in the F distribution is both very complex and very interesting.

5. The t distribution strongly resembles the standard normal distribution. However, the tails of the t distribution are thicker (or higher), and the t distribution is not as tall. The larger the degrees of freedom for a t distribution, the more closely it resembles the standard normal distribution. If we could have an infinitely large number of degrees of freedom, then the t distribution would be exactly the standard normal distribution.

6. If the normal approximation to the binomial distribution is used for inference procedures, it is vital that the assumption that the smaller of $n\pi$ and $n(1 - \pi)$ is at least 5 is satisfied. If it is not, the results of such an inference procedure could be wrong.

8.6 Lab Exercises

8.6.1 Critical Values

For the exercises in this section, choose Critical Values from the Z,t, Chi-square,F submenu.

1. Consider the case in which Type is Two-Tailed, Distribution is Z, and alpha = 0.10. What will the critical values be? Run the lab. Were you correct?

2. What is the value of alpha for the right-tailed critical value 1.96 from the standard normal (Z) distribution?

3   Run the lab for the t distribution, selecting each of Right-Tailed and Left-Tailed under Type, for the following combinations of degrees of freedom and α. Compare the t critical values to each other and to the Z critical values for the same values of alpha. What is happening to the t critical values as the values of alpha increase? What is happening to the closeness of t and Z as degrees of freedom increase?
   (a) df_1 = 5, alpha = 0.01, 0.05, and 0.10
   (b) df_1 = 25, alpha = 0.01, 0.05, and 0.10
   (c) df_1 = 50, alpha = 0.01, 0.05, and 0.10

4   Run the lab for the Chi-square distribution under the following conditions for each of the three choices under Type. Print each graph. If df_1 stays the same, what happens to the critical values as the value of alpha increases? For each value of alpha, compare the appearances of the tail areas as the degrees of freedom increase.
   (a) df_1 = 5, alpha = 0.01, 0.05, and 0.10
   (b) df_1 = 25, alpha = 0.01, 0.05, and 0.10
   (c) df_1 = 50, alpha = 0.01, 0.05, and 0.10

5   Run the lab for the F distribution under the following conditions for each of the three choices under Type. Print each graph. If df_1 and df_2 stay the same, what happens to the critical values as the value of alpha increases? For each value of alpha, compare the appearances of the tail areas as the degrees of freedom increase.
   (a) df_1 = 5, df_2 = 5, alpha = 0.01, 0.05, and 0.10
   (b) df_1 = 25, df_2 = 25, alpha = 0.01, 0.05, and 0.10
   (c) df_1 = 50, df_2 = 50, alpha = 0.01, 0.05, and 0.10

8.6.2 Normal Curves

For the exercises in this section, choose Normal Curves from the Z, t, Chi-square, F submenu.

6   Click on Reset, click on mu + five times, and print the resulting graph. What is the mean of the resulting distribution? What is the variance? How did increasing the mean affect the location of the distribution? What about the height and width of the distribution?

7   Click on Reset, click on mu − five times, and print the resulting graph. What is the mean of the resulting distribution? What is the variance? How did decreasing the mean affect the location of the distribution? What about the height and width of the distribution?

8   Click on Reset, click on var + three times, and print the resulting graph. What is the mean of the resulting distribution? What is the variance? How did increasing the variance affect the location of the distribution? What about the height and width of the distribution?

9 Click on `Reset`, click on `var -` three times, and print the resulting graph. What is the mean of the resulting distribution? What is the variance? How did decreasing the variance affect the location of the distribution? What about the height and width of the distribution?

10 Click on `Reset`, click on `mu +` three times, and click on `var +` three times. Print the resulting graph. How does the distribution change as the mean increases? What about as the variance increases?

11 Click on `Reset`, click on `mu -` three times, and click on `var -` three times. Print the resulting graph. How does the distribution change as the mean decreases? What about as the variance decreases?

12 Click on `Reset`, click on `mu +` three times, and click on `var -` three times. Print the resulting graph. How does the distribution change as the mean increases? What about as the variance decreases?

13 Click on `Reset`, click on `mu -` three times, and click on `var +` three times. Print the resulting graph. How does the distribution change as the mean decreases? What about as the variance increases?

14 Click on `Reset`. Beginning with `mu +`, click on `mu +` and then `var +` three times each. Print the resulting graph. Describe the change that is occurring in the distribution as the mean and the variance increase. How does the distribution change in terms of location and shape as the mean increases? How do the location and shape of the distribution change as the variance increases?

15 Click on `Reset`. Beginning with `mu +`, click on `mu +` and then `var -` three times each. Print the resulting graph. Describe the change that is occurring in the distribution as the mean and the variance change. How does the distribution change in terms of location and shape as the mean increases? How do the location and shape of the distribution change as the variance decreases?

16 Click on `Reset`. Beginning with `mu -`, click on `mu -` and then `var +` three times each. Print the resulting graph. Describe the change that is occurring in the distribution as the mean and the variance change. How does the distribution change in terms of location and shape as the mean decreases? How do the location and shape of the distribution change as the variance increases?

17 Click on `Reset`. Beginning with `mu -`, click on `mu -` and then `var -` three times each. Print the resulting graph. Describe the change that is occurring in the distribution as the mean and the variance decrease. How does the distribution change in terms of location and shape as the mean decreases? How do the location and shape of the distribution change as the variance decreases?

8.6.3 Chi-square Curves

For the exercises in this section, choose `Chi-square Curves` from the `Z,t, Chi-square,F` submenu.

18 Click on `df +` successively until `df = 100`. Describe, for each curve, what is happening to the shape of the new χ^2 distribution by answering the following questions. Print the final graph.

(a) What are the degrees of freedom for the new χ^2 distribution? What is the numerical value of the mean of the new χ^2 distribution? Where is the center of the new χ^2 distribution?

(b) Is the new distribution wider or narrower than the previous distributions? What is the numerical value of the variance of the new χ^2 distribution? How is this value related to the value of the variances of the previous distributions?

(c) Is the new χ^2 distribution more or less skewed than the previous distributions? At what degrees of freedom does the distribution begin to appear more symmetrical than skewed? What family of distributions do you think the χ^2 distribution would eventually look like if we let the degrees of freedom get large enough?

19 Let Inc = 90. Click on df + so that the resulting χ^2 distribution has df = 100. Now change Inc back to 10, and click on df - until df returns to 10. Answer the following questions for each new graph you generate.

(a) What are the degrees of freedom for the new χ^2 distribution? What is the numerical value of the mean of the new χ^2 distribution. Where is its center?

(b) Is the new distribution wider or narrower than the previous distributions? What is the numerical value of the variance of the new χ^2 distribution? How is this value related to the value of the variances of the previous distributions?

(c) Is the new χ^2 distribution more or less skewed than the previous distributions? At what degrees of freedom does the distribution begin to appear more skewed than symmetrical? What family of distributions do you think the χ^2 distribution would resemble for small degrees of freedom? (*Hint:* See the *How Are Populations Distributed?* Lab for help in answering this final question.)

8.6.4 F Curves

For the exercises in this section, choose F Curves from the Z, t, Chi-square, F submenu.

20 Click on Reset. Let Inc = 10, and click on df1 + until numerator degrees of freedom v_1 = 50. Print the resulting graph. For each of the five distributions, calculate the mean and the variance using the expressions given in Subsection 8.4.4. How is the mean affected by the increase in numerator degrees of freedom? What about the variance? How are these properties reflected in the changing appearance of the distribution?

21 Click on Reset. Let Inc = 90, and click on df1 + so that numerator degrees of freedom v_1 = 100. Now change Inc to 10, and click on df - until v_1 = 50. Print the resulting graph. For each of the five distributions, calculate the mean and the variance using the expressions given in Subsection 8.4.4. How is the mean affected by the decrease in numerator degrees of freedom? What about the variance? How are these properties reflected in the changing appearance of the distribution?

22 Click on Reset. Let Inc = 15, and click on df2 + once. Then change Inc to 25, and click on df2 + until the denominator degrees of freedom are df_2 = 200. Print the resulting graph. For each of the distributions, calculate the mean and

the variance. As the denominator degrees of freedom get larger and larger, what number is the mean approaching? What about the variance? How is this number related to the numerator degrees of freedom?

23 Click on Reset. Let Inc = 5. Beginning with df1 +, alternate clicking on df1 + and df2 + until you have clicked on them both five times. Print the resulting graph. For each curve, calculate the mean and the variance. What do you observe in the changing behavior of the location and shape of the curve? In numerical terms, how do the changing values of the mean and the variance reflect these behaviors?

8.6.5 *t* Converging to Z

For the exercises in this section, choose t Converging to Z from the Z, t, Chi-square, F submenu.

24 Let Xmin = −4, Xmax = −2, and df max = 5. Run the lab. What part of the distribution are you observing? What is happening as the degrees of freedom of the *t* distribution are increasing?

25 Let Xmin = 2, Xmax = 4, and df max = 20. Run the lab. What part of the distribution are you observing? What is happening as the degrees of freedom of the *t* distribution are increasing?

26 Let Xmin = −1, Xmax = 1, and df max = 20. Run the lab. What part of the distribution are you observing? What is happening as the degrees of freedom of the *t* distribution are increasing?

27 Let Xmin = −4, Xmax = −2, and df max = 20. Run the lab. What part of the distribution are you observing? What is happening as the degrees of freedom of the *t* distribution are increasing?

28 For the following list of values for Xmin and Xmax, run the lab for df max = 4, 15, 25, and 50. What part of the distribution are you observing? What is happening to the *t* distribution as degrees of freedom increase?

(a) Xmin = −4, Xmax = 4   (b) Xmin = −1, Xmax = 1
(c) Xmin = 2, Xmax = 4   (d) Xmin = −4, Xmax = −2

8.6.6 Normal Approximation to Binomial

For the exercises in this section, choose Normal Approximation to Binomial from the Z, t, Chi-square, F submenu.

29 Run the lab for each of n = 7, 10, and 20 with pi = 0.50. Print each graph. For which combinations of n and pi would you use the normal approximation to the binomial distribution for inference procedures? For which combinations would you not use the normal approximation, and why?

30 Run the lab for n = 20 with each of pi = 0.20, 0.50, and 0.80. Print each graph. For which combinations of n and pi would you use the normal approximation to the binomial distribution for inference procedures? For which combinations would you not use the normal approximation, and why?

9

Sampling Distributions

In the Z, t, Chi-square, and F Lab, we studied the features of four common statistical distributions. We saw that each distribution has its own unique shape and set of properties. It turns out that many of the statistics you will study in your introductory statistics course have one of these four distributions as a sampling distribution. In this lab, we will study five such statistics; the Z, one-sample t, two-sample t, χ^2, and F.

9.1 Introduction

The *sampling distribution* of a statistic is the distribution of the many possible values of the statistic that could be observed depending on which random sample is chosen from the population. The *Random Sampling* and *Central Limit Theorem* Labs introduced the idea of a sampling distribution. In the Z, t, Chi-square, and F Lab, we studied four common sampling distributions. We now will examine the closeness of the theoretical distributions to histograms of many observed values of a statistic. For purposes of statistical inference, the five statistics we will study may classified in the following manner:

1. One-sample inference procedures on the population mean μ:
   (a) *Case 1*: The Z statistic
   (b) *Case 2*: The (one-sample) t statistic
2. *Case 3*: The two-sample inference procedure on the difference between two population means $\mu_1 - \mu_2$; two-sample t statistic

3. *Case 4*: The one-sample inference procedure on the population variance σ^2; the χ^2 statistic
4. *Case 5*: The two-sample inference procedure on the ratio of two populations' variances σ_1^2/σ_2^2; the F statistic

9.1.1 Some Basic Ideas

In a statistical study, a sample of size n is randomly selected, and from that sample, a statistic is calculated. Many possible samples of size n could be selected from the population, each of which could yield a different value for the statistic. The sampling distribution is the pattern of how likely these different possible values for the statistic are to occur. The term *sampling* is included to emphasize the fact that the value of the statistic is entirely dependent on which sample is chosen. Knowing the sampling distribution of a statistic allows us to use that statistic to make decisions about aspects of the population. Exactly how the sampling distribution is used to accomplish this task is studied in the *Interpreting Confidence Intervals, Tests of Significance, Level of Significance of a Test*, and *Power of a Test* Labs (Chapters 11, 13, 14, and 16, respectively).

In this lab, we will study five different transformed statistics. Certain minimal conditions must be satisfied so that these statistics have the claimed sampling distribution. In our classification, for each of the one-sample procedures (cases 1, 2, and 4), we assume that a sample of size n is randomly selected from a normally distributed population with mean μ and variance σ^2. If the population is not normally distributed, then the sample size n must be "large." For the two-sample procedures (cases 3 and 5), we assume that samples of sizes n_1 and n_2 are randomly and independently selected from two independent, normally distributed populations. The mean and variance of the first population are μ_1 and σ_1^2, respectively. Similarly, for the second population, the mean is μ_2 and the variance is σ_2^2. As in the one-sample case, if the populations are not normally distributed, the sample sizes must be large. Three population distributions are considered in this lab. Expressions for the mean and variance of each of these population distributions are given in the *How Are Populations Distributed?* Lab.

If our basic assumptions are satisfied, then we have the following formulations for the transformed statistics and their sampling distributions.

1. *The Z statistic*: In the *Central Limit Theorem* Lab, we learned that if the population is normally distributed, then the sampling distribution of $\bar{X}$ is

$$N\left(\mu, \frac{\sigma^2}{n}\right).$$

If the population is not normally distributed but the sample size n is large, then the sampling distribution of $\bar{X}$ is still approximately normal. Moreover, the larger n becomes, the closer the sampling distribution gets to a normal distribution. Therefore, if the assumptions are satisfied, the sampling distribution of

$$Z = \frac{\bar{X} - \mu}{\sigma/\sqrt{n}}$$

is a $N(0,1)$ or a Z distribution.

2. *The one-sample t statistic*: The Z statistic requires that the value of the population variance σ^2 be known. If it is not, it must be estimated using s^2. The resulting transformed statistic is the t statistic:

$$t = \frac{\bar{X} - \mu}{s/\sqrt{n}}.$$

Because σ^2 must be estimated, the variability in the values we could observe for our test statistic is increased. This is because two different samples having the same value of $\bar{X}$ may have different values for s^2. Recall that the t distribution was shorter and had higher tails than the standard normal distribution. This difference in shapes means that the t distribution has a larger variance than the standard normal distribution, and so it is capable of explaining the added variability introduced by the estimation of σ^2. Thus, if the population being sampled from has a N(μ, σ^2) distribution, then the t statistic has a t_{n-1} distribution (the notation t_ν is explained in the *Z, t, Chi-square, and F* Lab). If n is large enough, this can be approximated by the N(0,1) distribution, because for large degrees of freedom, the t distribution converges to the N(0,1) distribution.

3. *The two-sample t statistic*: When working with two independent populations, a common approach is to compare the two means of the populations. The two-sample t statistic does this very well. An additional assumption when working with this statistic is that the two population variances σ_1^2 and σ_2^2 are equal. This does not imply that we know their value, but only that we know their value is equal. By making this assumption, we can estimate the common value of the population variances using a statistic called the *pooled variance*:

$$s_p^2 = \frac{(n_1 - 1)s_1^2 + (n_2 - 1)s_2^2}{n_1 + n_2 - 2}.$$

If we can make this final assumption, then the two-sample t statistic given by

$$t = \frac{(\bar{X}_1 - \bar{X}_2) - (\mu_1 - \mu_2)}{\sqrt{s_p^2\left(\frac{1}{n_1} + \frac{1}{n_2}\right)}}$$

has the t distribution with $\nu = n_1 + n_2 - 2$ degrees of freedom.

4. *The χ^2 statistic*: The χ^2 statistic is typically used for making inferences about the population variance σ^2. It arises in many ways, but most commonly as the sum of squares of ν independent standard normal random variables. In any case, if the population distribution being sampled has a normal distribution, or if the sample size n is large, the χ^2 statistic

$$\chi^2 = (n-1)\frac{s^2}{\sigma^2}$$

has a χ^2_{n-1} distribution.

5 *The F statistic*: When comparing the variances of two independent normal distributions, the F statistic is appropriate. It arises as the ratio of two independent χ^2 statistics divided by their respective degrees of freedom:

$$F = \frac{s_1^2/\sigma_1^2}{s_2^2/\sigma_2^2}.$$

Under the previously stated assumptions, this ratio has an F_{n_1-1,n_2-1} distribution.

9.2 Objectives

1 Observe the closeness of the observed sampling distribution and theoretical sampling distribution of a statistic.
2 Learn when the theoretical sampling distribution is appropriate for use in statistical inference.
3 Understand the effects of violating assumptions on the appropriateness of using the theoretical sampling distribution for purposes of statistical inference.

9.3 Description

To run the lab, you will need to choose which statistic to study, which population to sample from, and what sample size to use. Five hundred samples (or pairs of samples) of size n (or of size n_1 and n_2) will be generated from the specified population. For each sample, the transformed statistic will be calculated. A histogram of the 500 values of the statistic will be drawn, and the theoretical curve of the appropriate sampling distribution will be superimposed on the histogram for a graphical comparison. In addition, a table of the percentiles of the histogram and the theoretical curve is given so that you can numerically compare how closely the two agree.

Start the lab by highlighting and clicking on Sampling Distributions from the Labs menu. When the lab begins, the dialog box shown in Figure 9.1 opens, which briefly describes the lab. Below the description are four smaller boxes requiring information.

1 n_1: The value given in the box to the immediate right of n_1 represents the size of the samples chosen in the study if Statistic is Z, One-sample-t, or Chi-square. If the choice under Statistic is either Two-sample-t or F, n_1 represents the size of the samples chosen from the first population. The initial value of n_1 is 5. The values of n_1 should be integers and must be at least 5.

2 n_2: The value given in the box to the immediate right of n_2 represents the size of the samples chosen from the second population if Two-sample-t or F is specified in Statistic. If Statistic is Z, One-sample-t, or Chi-square, the value of n_2 is ignored. The initial value of n_2 is 5. The values of n_2 should be integers and must be at least 5.

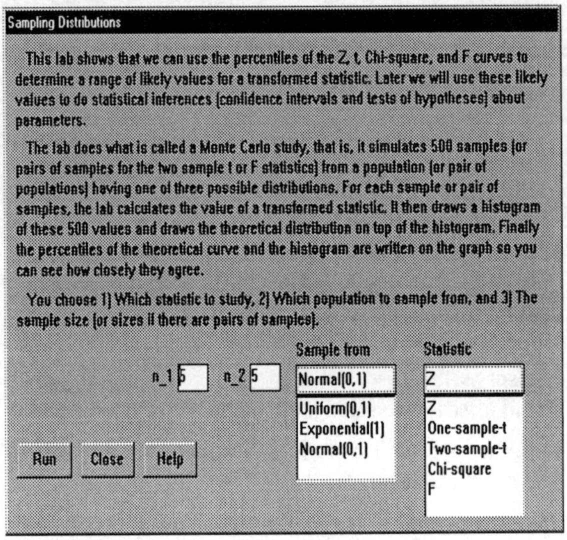

FIGURE 9.1
Dialog box for the *Sampling Distributions* Lab.

3. `Sample from`: The distribution chosen under `Sample from` specifies the distribution of the population from which samples are generated. There are three distributions that may be selected: `Normal(0,1)`, `Exponential(1)`, or `Uniform(0,1)`. For more information on these distributions, see the *How Are Populations Distributed?* Lab. The initial distribution is `Normal(0,1)`. The choices for `Sample from` are limited to those shown.

4. `Statistic`: The statistic chosen here corresponds to one of the five statistics discussed in Subsection 9.1.1. The initial statistic specified for study is the Z statistic. The choices of `Statistic` are limited to those shown.

In the bottom left-hand corner of the dialog box are three buttons.

1. Run: Clicking on this button begins the lab. A graphics window will open briefly, containing a bar that fills with yellow as the 500 samples are generated. After the samples are selected and computations are completed, a graph similar to the one shown in Figure 9.2 will appear. Two methods of comparison are used:

   (a) For visual comparison, a histogram of the 500 values of the transformed statistic is displayed with the curve of the theoretical sampling distribution superimposed.

   (b) To the right of this graph of distributions is a table containing percentiles from the theoretical distribution and the histogram. In this table, the first column represents the area under the theoretical distribution to the left of the number adjacent to it in the next column to the right. Thus, in Figure 9.2, 0.025 is the area to the left of -1.96 for a true standard normal distribution. A number having area $A/100$ to the left of it is called the *Ath percentile*. Therefore, -1.96 is the 2.5th percentile of a standard normal distribution. The third column lists the same percentiles but for the 500 values of the statistic. Thus, in Figure 9.2, 90% of the values of Z are less than or equal to 1.20 while 90% of the elements in a Z population are less than or equal to 1.28. The last

FIGURE 9.2
Histogram of 500 Z statistics from samples of size $n = 5$ from a N(0,1) distribution with the Z curve superimposed.

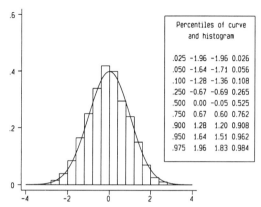

column gives the proportion of the 500 values of the statistic that are less than or equal to the percentile of the theoretical curve. In Figure 9.2, 96.2% of the 500 values of Z are less than or equal to 1.64 while 95% of the values in a Z population are less than or equal to 1.64.

2 Close: Clicking on this button closes the lab and returns you to the StatConcepts menus.

3 Help: Clicking on this button opens a help window containing information about the lab.

9.4 Guided Tour of the Lab

The tour you are about to take will explore many fascinating sights! Odd as it may sound, we'll begin at the Z statistic and continue on through the F statistic. By the end of this tour, your appreciation of sampling distributions will have deepened dramatically.

9.4.1 The Sampling Distribution of the Z Statistic

Earlier, we claimed that if either the parent population had a normal distribution or the sample size was sufficiently large, then the Z statistic, given by

$$Z = \frac{\bar{X} - \mu}{\sigma/\sqrt{n}},$$

had a standard normal (or N(0,1)) distribution. We are now going to examine the validity of that claim. We'll begin by looking at the distribution of Z for samples chosen from a N(0,1) distribution. Then we'll see what happens if the parent population does not have a normal distribution. In particular, we'll try to determine just how large n must be before concluding that the normal distribution for the sampling distribution of $\bar{X}$ is reliable.

Make sure `Statistic` is Z, and let `n_1` = 5, and `Sample from` be `Normal(0,1)`. Then run the lab. When you run the lab, 500 samples from an N(0,1) distribution are being generated, and for each of those 500 samples, the value of the Z statistic is being calculated. When the computations are complete, the resulting graph will look somewhat like the one shown in Figure 9.2. It shouldn't look exactly the same, because each time the lab is run, a new set of 500 samples is generated.

In this graph, the histogram is of the 500 values of the Z statistic. The red curve is a graph of the N(0,1) (or standard normal) distribution. Notice how closely the two match. Both are symmetrical around zero, and both have a bell-shaped appearance. Graphically, the two distributions appear to agree closely, but we know that looks can be deceiving. So how closely do the two agree numerically?

To the right of the picture of the distributions is a table that compares percentiles of the theoretical normal distribution to the histogram. The first column of numbers (which includes 0.025, 0.050, ..., 0.975) are areas to the left of the numbers in the second column (−1.96, −1.64, ..., 1.96) for the theoretical (standard normal) distribution. Thus, for example, for a N(0,1) distribution, the area to the left of −1.96 is 0.025, the area to the left of 1.28 is 0.900, and so on. The numbers in the second column are the percentiles of the standard normal distribution; −1.96 is the 2.5th percentile, 1.28 is the 90th percentile, and so on.

The third column of numbers (−1.96, −1.71, ..., 1.83) are percentiles from the histogram, so that the area to the left of these numbers is the closest possible match from the histogram to the areas listed in the first column from the Z distribution. The numbers in the fourth column (0.026, 0.056, ..., 0.984) are the areas from the histogram to the left of the percentiles in the second column. Notice how closely both the two columns of percentiles and the two columns of areas agree. They are not exactly the same, but they are close. We have only a small number (500) of samples, so we shouldn't expect the columns to be exactly the same. But if we could generate an infinite number of samples, then the columns of percentiles (and of areas) would, in fact, be exactly the same.

Will increasing the sample size affect the closeness of the two distributions? Not in this case. Recall that if the parent population has a normal distribution (and in this case it does, because `Sample from` is `Normal(0,1)`), then the distribution of $\bar{X}$ is exactly normal, so the distribution of Z is exactly N(0,1), regardless of the sample size.

What happens if the parent population does not have a normal distribution? Change `Sample from` to `Uniform(0,1)` and run the lab to find out. Figure 9.3 shows one possible result. Examination of the graphs and the numerical table seems to indicate that these two distributions are in close agreement. If you reflect back on the *Central Limit Theorem* Lab, this result should not surprise you. In that lab, we saw that for a uniformly distributed parent population, $n = 5$ was an adequately large sample size to justify the claim that the sampling distribution of $\bar{X}$ was normal. Because Z is merely a simple transformation of $\bar{X}$, the same holds true for Z.

Is $n = 5$ large enough when the parent population has an exponential(1) distribution? Based on the *Central Limit Theorem* Lab, we should suspect that it is not. Run the lab and see. Let `Sample from` be `Exponential(1)` and `n_1` = 5. Your result should look something like what is shown in Figure 9.4. No surprise here! The histogram of the 500 Z's is skewed. Graphically, we can tell that these two distributions

FIGURE 9.3
Histogram of 500 Z statistics from samples of size $n = 5$ from a Uniform(0,1) distribution with the Z curve superimposed.

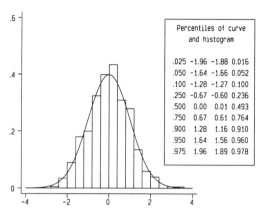

FIGURE 9.4
Histogram of 500 Z statistics from samples of size $n = 5$ from an Exponential(1) distribution with the Z curve superimposed.

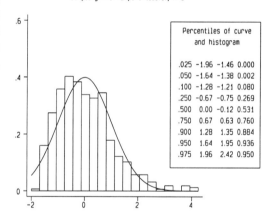

do not match, and the numerical comparison tells the same story, particularly in the left tail. The best we can do using the histogram is around the 93.6th percentile, and it is 1.95. So how do we justify using the standard normal distribution as the sampling distribution of Z when the parent population has an Exponential(1) distribution? The Central Limit Theorem tells us to increase the sample size and the distribution will become normal. So how large is large? In Figure 9.5, the value of n_1 was 60, and the Z distribution and histogram are in much closer agreement. Can n be smaller? In the chapter exercises, you will find out!

9.4.2 The Sampling Distribution of the One-sample t Statistic

An interesting phenomenon occurs when the value of σ^2 in the Z statistic is unknown and so must be estimated by s^2. When this happens, the distribution of the new statistic, the one-sample t statistic, is not a Z distribution any more. It is a t_{n-1} distribution. An important assumption here is that the parent distribution is from the

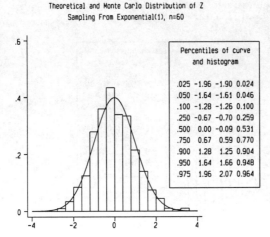

FIGURE 9.5
Histogram of 500 Z statistics from samples of size $n = 60$ from an Exponential(1) distribution with the Z curve superimposed.

normal family. In this subsection, we will examine the effect of changes in sample size, of non-normal parent distributions, and of a combination of both on the sampling distribution of the one-sample t.

If you haven't already done so, change Statistic to One-sample-t. Let n_1 = 5 and Sample from be Normal(0,1). Then run the lab. The distribution of a one-sample t statistic, based on a sample of size $n = 5$ from a normal distribution, is a t_4 distribution.

In Figure 9.6, the agreement between the histogram and the theoretical t_4 distribution appears to be pretty good, and numerically there is no reason to doubt this conclusion. Even around the center of the two distributions, where the graph seems to indicate that there might be a discrepancy, the numerical comparisons reveal that it is only a minor one. In this case, where all assumptions are satisfied, there is only one disadvantage that results from the need to estimate σ^2. Notice the range of the possible values on the horizontal axis. With such a small sample size, the one-sample t statistic

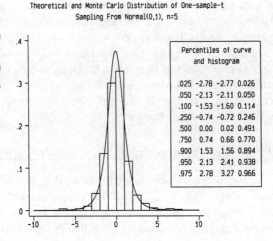

FIGURE 9.6
Histogram of 500 one-sample t statistics from samples of size $n = 5$ from a Normal(0,1) distribution with the t_4 curve superimposed.

has a chance of being between approximately −8 and 8. Granted, the extremes are highly unlikely, but they are likely enough that they did show up on the histogram.

Now increase n_1 to 10 and run the lab again. Because the sample size has changed, the degrees of freedom for the t distribution are $n − 1 = 9$, and so the theoretical sampling distribution of the one-sample t is now t_9. (Recall that changing the sample size had no effect on the sampling distribution of the Z statistic.) Again, the theoretical and Monte Carlo distributions agree closely, but now the values of the one-sample t are less variable—they only range between −5 and 5 (see Figure 9.7). This is because sample size, and hence the degrees of freedom for the t distribution, is increasing. The more information we get from our sample, the less variable are the possible responses. In other words, a larger sample size implies a more precise response.

Now run the lab for n_1 = 50. As we expected, the values are getting less and less variable (see Figure 9.8). One other behavior you might notice: The t distribution is looking more and more like the Z distribution. We actually observed this in the t *Converging to Z* Lab.

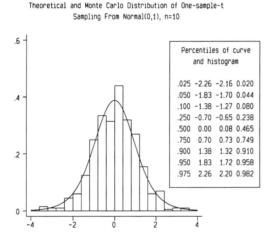

FIGURE 9.7
Histogram of 500 one-sample t statistics from samples of size $n = 10$ from a Normal(0,1) distribution with the t_9 curve superimposed.

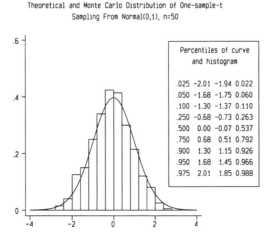

FIGURE 9.8
Histogram of 500 one-sample t statistics from samples of size $n = 50$ from a Normal(0,1) distribution with the t_{49} curve superimposed.

Earlier, we mentioned that the assumption about the parent distribution being normally distributed was of some importance. But how important? Well, we are about to see that this depends on which distribution the parent population follows. Change n_1 back to 5, and `Sample from` to `Uniform(0,1)`, and run the lab. One possible result is shown in Figure 9.9. Because the sample size is small, just as when the parent population was `Normal(0,1)`, the variability in the distributions is large. But there is more to notice here. Although the shapes of the distributions are similar, the percentiles give us reason for concern. The areas are comparable, but the corresponding percentiles are somewhat different. If the t_4 distribution was used in this case, our inference procedures will not be as reliable as we would like. We will study exactly how this affects inference results in the *Interpreting Confidence Intervals*, *Level of Significance of a Test*, and *Power of a Test* Labs (Chapters 11, 14, and 16, respectively).

Now change n_1 to 10 and run the lab again. The closeness of these new distributions, shown in Figure 9.10, is more reassuring. The variability in the possible

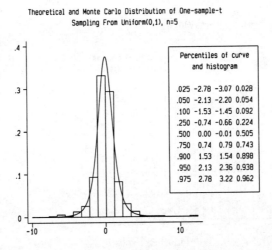

FIGURE 9.9
Histogram of 500 one-sample t statistics from samples of size $n = 5$ from a Uniform(0,1) distribution with the t_4 curve superimposed.

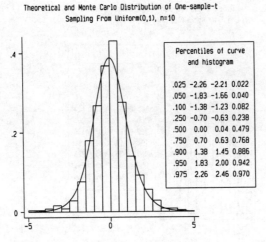

FIGURE 9.10
Histogram of 500 one-sample t statistics from samples of size $n = 10$ from a Uniform(0,1) distribution with the t_9 curve superimposed.

responses is reduced (because the sample size has increased). And the two distributions are also in closer agreement. Graphically, the theoretical curve and the Monte Carlo distribution appear similar. Numerically, they are more alike than when $n = 5$.

Increase the sample size once more to n_1 = 50. Figure 9.11 further confirms that the larger the sample size gets, the more reliable the t_{n-1} distribution becomes as a sampling distribution for the one-sample t statistic, even if the population is not normally distributed. In the case of a uniformly distributed parent population, a "large" sample size is somewhere between $n = 10$ and $n = 50$. In the chapter exercises, we will try to narrow this range.

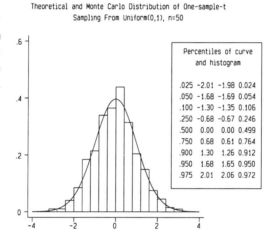

FIGURE **9.11**
Histogram of 500 one-sample t statistics from samples of size $n = 50$ from a Uniform(0,1) distribution with the t_{49} curve superimposed.

What happens to the closeness between the Monte Carlo distribution and the theoretical t_{n-1} distribution if the parent population is exponentially distributed? Figures 9.12, 9.13, and 9.14 show the results of running the lab for Sample from

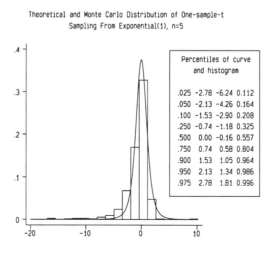

FIGURE **9.12**
Histogram of 500 one-sample t statistics from samples of size $n = 5$ from an Exponential(1) distribution with the t_4 curve superimposed.

FIGURE 9.13
Histogram of 500 one-sample t statistics from samples of size $n = 10$ from an Exponential(1) distribution with the t_9 curve superimposed.

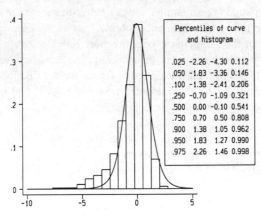

FIGURE 9.14
Histogram of 500 one-sample t statistics from samples of size $n = 50$ from an Exponential(1) distribution with the t_{49} curve superimposed.

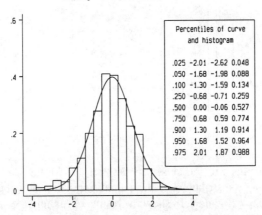

being Exponential(1) and n_1 = 5, 10, and 50, respectively. Notice the range of the values of the sample means. For n_1 = 5, the values range from around −20 to 10. The Monte Carlo distribution is skewed to the left, but the t_4 distribution is symmetrical around zero. This may be difficult to see because of the skewness of the Monte Carlo distribution. However, if you compare the t_4 distribution on this graph to the previous graphs containing t_4 distributions, you'll see that the theoretical curves are exactly the same. Thus, the graphs are not in agreement. Further, the differences in the numerical comparisons are alarming! From this, we can conclude that if the parent population is exponentially distributed, using a t_4 for the sampling distribution of the one-sample t statistic would be ill-advised.

When n_1 = 10, the Monte Carlo distribution ranges from −10 to 5 and is still skewed, but not as severely. The numbers in the table of percentiles and the areas are still quite disturbing. Therefore, we should be wary of relying on inferences using the t distribution for inferential purposes. For n_1 = 50, there is still some skewness, but it appears slight. An examination of the numerical quantities reveals that the two distributions are closer in agreement. It would most likely be acceptable for $n = 50$ to use the t_{49} distribution for purposes of statistical inference, but one should do so advisedly.

9.4.3 The Sampling Distribution of the Two-sample t Statistic

The two-sample t statistic is used in statistical inference for comparing the means of two independent, normally distributed populations with equal variances. If samples are randomly and independently chosen, then the sampling distribution of the two-sample t statistic is $t_{n_1+n_2-2}$.

If you haven't already done so, change Statistic to Two-sample-t. Let n_1 and n_2 = 5, and the distribution under Sample from be Normal(0,1). The results are shown in Figure 9.15. For these sample sizes, the degrees of freedom for the t distribution are $n_1 + n_2 - 2 = 5 + 5 - 2 = 8$. As we expect, because the assumptions are satisfied, the theoretical and Monte Carlo distributions appear very similar both graphically and numerically.

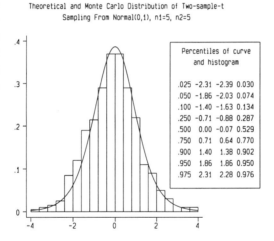

FIGURE 9.15
Histogram of 500 two-sample t statistics from samples of sizes $n_1 = n_2 = 5$ from Normal(0,1) populations with the t_8 curve superimposed.

If we increase the sample sizes so that both n_1 and n_2 = 10, as shown in Figure 9.16, the only noticeable difference between the two graphs is that the tails of the t_{18} distribution are not as "thick" as the tails of the t_8 distribution. In both cases, the Monte Carlo distributions and the theoretical t distributions agree nicely.

We now need to see what happens to this agreement if the parent distributions are not normal. We begin by considering the Uniform(0,1) distribution. Based on what we have already observed for uniform parent distributions, we expect the Monte Carlo and theoretical distributions to come very close to agreement for relatively small sample sizes. Begin with n_1 and n_2 = 5, and be sure Sample from is Uniform(0,1). Run the lab. As Figure 9.17 shows, even for these small sample sizes, the Monte Carlo and theoretical t_8 distributions agree closely both graphically and numerically. But will this agreement hold if the parent population has an exponential distribution?

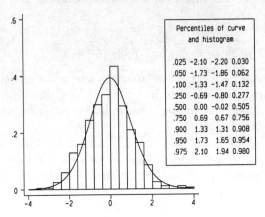

FIGURE 9.16
Histogram of 500 two-sample t statistics from samples of sizes $n_1 = n_2 = 10$ from Normal(0,1) populations with the t_{18} curve superimposed.

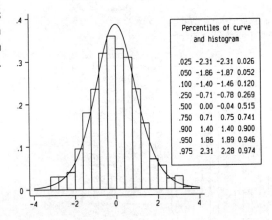

FIGURE 9.17
Histogram of 500 two-sample t statistics from samples of sizes $n_1 = n_2 = 5$ from Uniform(0,1) populations with the t_8 curve superimposed.

Change Sample from to Exponential(1), and run the lab. You will see a result similar to what is shown in Figure 9.18. Although the results don't look too bad, they also aren't nearly as good as when the parent distributions were uniform.

Now increase the sample sizes to n_1 and n_2 = 15. The result is shown in Figure 9.19. The Monte Carlo and theoretical t_{28} distributions are much closer to each other both graphically and numerically.

One final note about the two-sample t statistic: It appears to require a smaller sample size for the Monte Carlo distribution to come close to the theoretical curve. However, recall that the expression for degrees of freedom is $v = n_1 + n_2 - 2$, so both sample sizes are involved in determining the closeness of the Monte Carlo and theoretical distributions. Seen in this light, there is actually very little difference in the total number of elements chosen for the one-sample and two-sample t statistics before the Monte Carlo and theoretical t distributions agree.

FIGURE 9.18
Histogram of 500 two-sample t statistics from samples of sizes $n_1 = n_2 = 5$ from Exponential(1) populations with the t_8 curve superimposed.

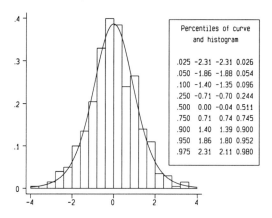

FIGURE 9.19
Histogram of 500 two-sample t statistics from samples of sizes $n_1 = n_2 = 15$ from Exponential(1) populations with the t_{28} curve superimposed.

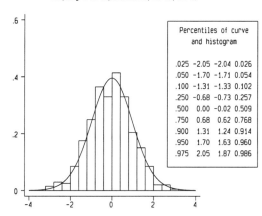

9.4.4 The Sampling Distribution of the χ^2 Statistic

The χ^2 statistic is used for purposes of statistical inference on the population variance σ^2. If a sample of size n is drawn from a normal parent population, the sampling distribution of the χ^2 statistic is χ^2_{n-1}. In this subsection, we'll decide if the same principles that applied to the sampling distribution of the Z and t statistics apply to the sampling distribution of a χ^2 statistic. Before we proceed, recall that the χ^2_ν distribution was skewed to the right and began at zero (unlike the Z and t_ν distributions, which were symmetrical and centered at zero).

First, let's examine the histogram of the 500 χ^2 statistics and the theoretical curve of the χ^2 sampling distribution when the parent population is Normal(0,1). Let n_1 = 5, and if you have not already done so, change Statistic to Chi-square. The result is shown in Figure 9.20. You should not be surprised by the close agreement of the two distributions. The assumption of the normality of the parent population

FIGURE 9.20

Histogram of 500 χ^2 statistics from samples of size $n_1 = 5$ from a Normal(0,1) population with the χ_4^2 curve superimposed.

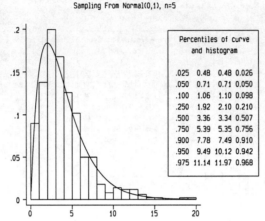

distribution was satisfied. In this case, just as in the Z and t cases, if the assumptions are satisfied, the relationship is exact; that is, the sampling distribution of the χ^2 statistic is exactly χ_4^2.

What happens to the sampling distribution of the χ^2 statistic if the parent distribution is not a member of the normal family? For the same value of n_1, run the lab for a Uniform(0,1) parent distribution. One possible result is shown in Figure 9.21. If you remember what occurred in this case for the Z and t statistics, the result will not be at all what you might expect. The agreement between the Monte Carlo distribution and the theoretical χ_4^2 distribution is not very good. In the graph, it is fairly obvious that the Monte Carlo distribution is not as skewed as the χ_4^2 distribution. A numerical comparison of the two distributions confirms that using the χ_4^2 distribution when the sample size is $n = 5$ from a uniformly distributed parent could be disastrous.

In the Z and t cases, increasing the sample size helped to improve the closeness of the two distributions very quickly for the uniform parent distribution. To see if this

FIGURE 9.21

Histogram of 500 χ^2 statistics from samples of size $n_1 = 5$ from a Uniform(0,1) population with the χ_4^2 curve superimposed.

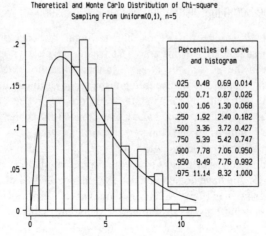

is so in the χ^2 case, change n_1 to 20 and run the lab again. The result is shown in Figure 9.22. Note that the graphs still are not in close agreement. In fact, the Monte Carlo distribution more closely resembles a normal distribution with mean 19 than a χ^2_{19} distribution. Therefore, the center of the χ^2 distribution is correct (recall that the mean of a χ^2_ν distribution is its degrees of freedom), but the skewness does not appear appropriate. The numerical comparisons only stengthen our concerns. But be careful not to assume, based on the appearance of the graph, that the approximation is worse. The horizontal scales for the two graphs are different. Our conclusion: If you use the χ^2_ν distribution for inference on σ^2 when the parent population is uniform, then do so with caution.

Is this also true for an exponentially distributed parent population? Change n_1 back to 5, and choose Exponential(1) for Sample from. Figure 9.23 shows one possible result. Graphically, things don't look so good, and numerically, they look even worse. Will increasing sample size help?

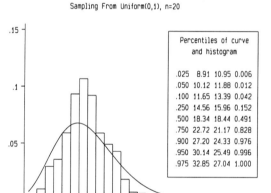

FIGURE **9.22** Histogram of 500 χ^2 statistics from samples of size $n_1 = 20$ from a Uniform(0,1) population with the χ^2_{19} curve superimposed.

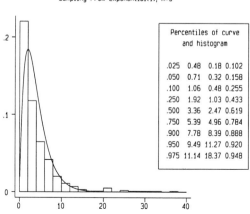

FIGURE **9.23** Histogram of 500 χ^2 statistics from samples of size $n_1 = 5$ from an Exponential(1) population with the χ^2_4 curve superimposed.

Now change n_1 to 20 and run the lab again. Unfortunately, Figure 9.24 does little to reassure us that these two distributions will agree anytime in the near future if we continue to increase the sample size.

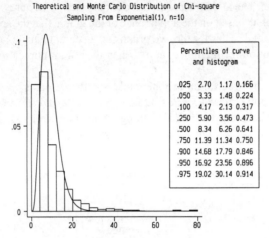

FIGURE 9.24
Histogram of 500 χ^2 statistics from samples of size $n_1 = 20$ from an Exponential(1) population with the χ^2_{19} curve superimposed.

We have just witnessed that larger sample sizes may help more in some cases than in others. If we continued to choose larger sample sizes, the agreement between the χ^2 sampling distribution and the theoretical χ^2_{n-1} distribution would become quite good. But, as we have seen, the sample size must be quite large. Therefore, in the case of the χ^2 statistic, for small and moderate sample sizes, the parent population having a normal distribution is much more critical than it is for the Z or t statistics.

9.4.5 The Sampling Distribution of the F Statistic

The F statistic is used for comparing two population variances. If the parent populations are independently and normally distributed, then the F statistic has an F_{n_1-1, n_2-1} distribution. As we have done in previous cases, we will examine the closeness of the Monte Carlo distribution and the theoretical F_{n_1-1, n_2-1} distribution under a variety of circumstances, including a normally distributed parent population, a uniform parent population, and the exponential parent population. Before you continue, you may want to review the properties of the F_{v_1, v_2} distribution from the F Curves Lab.

Change Statistic to F, choose Normal(0,1) for Sample from, and let both n_1 and n_2 = 10. Then run the lab. The results are shown in Figure 9.25. As you might expect, the Monte Carlo and $F_{9,9}$ distributions are in very close agreement. The assumptions are satisfied, and the relationships are exact.

Now we will see what happens when the parent population has a distribution from the family of uniform distributions. Change Sample from to Uniform(0,1) and run the lab again. In Figure 9.26, we see that, just as in the χ^2 case, the two distributions are not very similar. The graphical and numerical comparisons imply that using the $F_{9,9}$ distribution in this case would be dangerous.

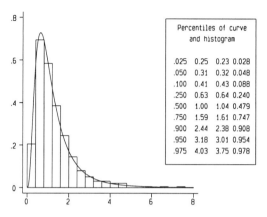

FIGURE **9.25**
Histogram of 500 F statistics from samples of sizes $n_1 = n_2 = 10$ from Normal(0,1) populations with the $F_{9,9}$ curve superimposed.

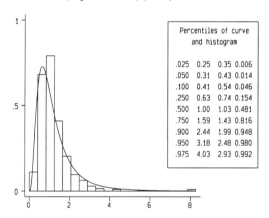

FIGURE **9.26**
Histogram of 500 F statistics from samples of sizes $n_1 = n_2 = 10$ from Uniform(0,1) populations with the $F_{9,9}$ curve superimposed.

Now change both n_1 and n_2 to 20 to see if increasing sample size improves the closeness of the two distributions. Figure 9.27 shows the results. At first glance, the $F_{19,19}$ distribution actually seems to be doing a poorer job of estimating the sampling distribution of the F statistic. However, this is an illusion, because the scales on the horizontal axis are different. A numerical comparison reveals that the distributions are actually closer for larger sample sizes.

Now to examine the goodness of the F_{n_1-1,n_2-1} distribution for an exponentially distributed parent. Change n_1 and n_2 back to 10, and change Sample from to Exponential(1). Then run the lab. Figure 9.28 shows the results. The Monte Carlo distribution is so skewed that it is difficult to tell graphically if these two compare or not. But a quick examination of the percentiles and areas tells us they do not!

How effective is an increase in the sample size? Let n_1 and n_2 = 20, and run the lab to find out. The results are shown in Figure 9.29. Again, graphically, it is difficult to tell if the theoretical $F_{19,19}$ curve is doing a better job of estimating the

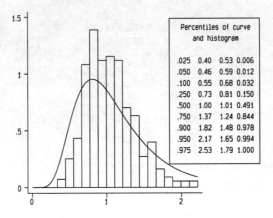

FIGURE **9.27**
Histogram of 500 F statistics from samples of sizes $n_1 = n_2 = 20$ from Uniform(0,1) populations with the $F_{19,19}$ curve superimposed.

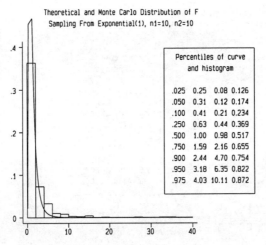

FIGURE **9.28**
Histogram of 500 F statistics from samples of sizes $n_1 = n_2 = 10$ from Exponential(1) populations with the $F_{9,9}$ curve superimposed.

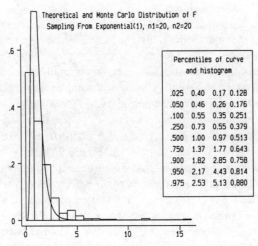

FIGURE **9.29**
Histogram of 500 F statistics from samples of sizes $n_1 = n_2 = 20$ from Exponential(1) populations with the $F_{19,19}$ curve superimposed.

sampling distribution of the F statistic. But the numerical comparisons tell us that it is indeed improving.

Recall that the F statistic is a function of two χ^2 statistics. This explains why we witnessed similar results for the sampling distribution of the F statistic and the χ^2 statistic. Specifically:

1. If the parent population is `Normal(0,1)`, the histogram and theoretical curve appear very similar, and the theoretical percentiles and percentiles from the histogram are very close to one another for approximately the same areas.

2. However, because the F distribution is skewed and begins at zero (like the χ^2 distribution), the assumption that the parent population is normally distributed plays a larger role in a reliable use of these sampling distributions. Small to moderate sample sizes will not suffice in this case. For the approximation to be reliable, the sample sizes will have to be larger.

9.5 Summary

We saw many sights on this tour:

1. The sampling distribution of a statistic is the theoretical distribution that describes the pattern of likelihood for values of the statistics calculated from all possible samples from the parent population. The term *sampling* is included to emphasize the fact that the value of the statistic is entirely dependent on which sample is selected from the population.

2. A Monte Carlo simulation study compares the actual behavior of many values of the statistic from different samples to the theoretical behavior.

3. For a normally distributed parent population, the Z statistic has a standard normal, or Z, distribution.

4. For a normally distributed parent population, the one-sample t statistic has a t_{n-1} distribution.

5. For normally distributed parent populations with equal variances, the two-sample t statistic has a $t_{n_1+n_2-2}$ distribution.

6. For these three cases, if the parent population was not normally distributed, increasing the sample size caused the sampling distribution to become more normal. We examined two specific cases:

    (a) If the parent population was uniformly distributed (symmetrical, but not normal), then a small to moderate sample size was enough to ensure that the sampling distribution was very close to the claimed theoretical curve.

    (b) If the parent population was exponentially distributed (skewed), then the sample size needed to be larger to justify using the theoretical sampling distribution for purposes of statistical inference.

7. For a normally distributed parent population, the χ^2 statistic has a χ^2_{n-1} distribution.

8 For normally distributed parent populations, the F statistic has a F_{n_1-1, n_2-1} distribution.

9 For these two distributions, the assumption that the parent population is normally distributed is more critical than it is in the case of the Z and t statistics.

9.6 Lab Exercises

9.6.1 The Sampling Distribution of the Z Statistic

In these exercises, the value of Statistic must be Z.

1 Run the lab for Sample from Normal(0,1), with n_1 = 25 and 100. Print both graphs. Does a larger sample size in the normal distribution seem to better describe the sampling distribution of Z than a smaller sample size? Explain your answer.

2 Let Sample from be Uniform(0,1). For each of the following values of n_1, run the lab and print the graphs. Comment on the closeness of the two distributions in each graph. How large does the sample size need to be so that using the N(0,1) distribution is a good approximation for the sampling distribution of Z?

(a) n_1 = 1   (b) n_1 = 3   (c) n_1 = 5   (d) n_1 = 7
(e) n_1 = 9   (f) n_1 = 15   (g) n_1 = 25   (h) n_1 = 30

3 Let Sample from be Exponential(1). For each of the following values of n_1, run the lab and print the graphs. Comment on the closeness of the two distributions in each graph. How large does the sample size need to be so that using the N(0,1) distribution is a good approximation for the sampling distribution of Z? How does this compare to your conclusions in the previous exercise when Sample from was Uniform(0,1)?

(a) n_1 = 1   (b) n_1 = 3   (c) n_1 = 5
(d) n_1 = 7   (e) n_1 = 9   (f) n_1 = 15
(g) n_1 = 25   (h) n_1 = 30   (i) n_1 = 50

9.6.2 The Sampling Distribution of the One-sample t Statistic

In these exercises, the value of Statistic must be One-sample-t.

4 Run the lab for Sample from Normal(0,1) for each of the sample sizes given below, and print the graphs. What is happening to the shape of the distribution as the sample size increases? Relate this to the increase in degrees of freedom. Is there another sampling distribution you could use for the sampling distribution of the one-sample t statistic for very large n?

(a) n_1 = 3   (b) n_1 = 5   (c) n_1 = 10
(d) n_1 = 25   (e) n_1 = 30   (f) n_1 = 50

5. Let Sample from be Normal(0,1), and n_1 = 100. Run the lab for Statistic being each of Z and One-sample-t. Print both graphs. How do the four distributions compare graphically and numerically?

6. Let Sample from be Uniform(0,1). For each of the following values of n_1, run the lab and print the graphs. Comment on the closeness of the two distributions in each graph. How large does the sample size need to be so that using the t_{n-1} distribution is a good approximation for the sampling distribution of the one-sample t statistic?

   (a) n_1 = 3   (b) n_1 = 5   (c) n_1 = 7
   (d) n_1 = 9   (e) n_1 = 11 (f) n_1 = 13
   (g) n_1 = 15 (h) n_1 = 17 (i) n_1 = 19

7. Let Sample from be Exponential(1). For each of the following values of n_1, run the lab and print the graphs. Comment on the closeness of the two distributions in each graph. How large does the sample size need to be so that using the t_{n-1} distribution is a good approximation for the sampling distribution of the one-sample t statistic? How does this compare with the previous exercise when Sample from was Uniform(0,1)?

   (a) n_1 = 5   (b) n_1 = 10   (c) n_1 = 15   (d) n_1 = 25
   (e) n_1 = 30   (f) n_1 = 35   (g) n_1 = 40

9.6.3 The Sampling Distribution of the Two-sample t Statistic

In these exercises, the value of Statistic must be Two-sample-t.

8. Run the lab for Sample from Normal(0,1) for each of the following sample sizes, and print the graphs. What is happening to the shape of the distribution as the sample size increases? Relate this to the increase in degrees of freedom. Is there another sampling distribution you could use for the sampling distribution of the two-sample t statistic for very large n?

   (a) n_1 = n_2 = 3   (b) n_1 = n_2 = 5   (c) n_1 = n_2 = 10
   (d) n_1 = n_2 = 25   (e) n_1 = n_2 = 30   (f) n_1 = n_2 = 50

9. Let Sample from be Normal(0,1), and n_1 = n_2 = 50. Run the lab for Statistic of both Z and Two-sample-t. Print both graphs. How do the four distributions compare graphically and numerically?

10. Let Sample from be Uniform(0,1). For each of the following values of n_1 and n_2, run the lab and print the graphs. Comment on the closeness of the two distributions in each graph. How large does the sample size need to be so that using the $t_{n_1+n_2-2}$ distribution is a good approximation for the sampling distribution of the two-sample t statistic?

    (a) n_1 = n_2 = 3   (b) n_1 = n_2 = 5   (c) n_1 = n_2 = 6
    (d) n_1 = n_2 = 7   (e) n_1 = n_2 = 8   (f) n_1 = n_2 = 9
    (g) n_1 = n_2 = 10

11 Let Sample from be Exponential(1). For each of the following values of n_1 and n_2, run the lab and print the graphs. Comment on the closeness of the two distributions in each graph. How large does the sample size need to be so that using the $t_{n_1+n_2-2}$ distribution is a good approximation for the sampling distribution of the two-sample t statistic? How does the rate of convergence compare to the previous exercise when Sample from was Uniform(0,1)?

(a) n_1 = n_2 = 5 (b) n_1 = n_2 = 10 (c) n_1 = n_2 = 15
(d) n_1 = n_2 = 25 (e) n_1 = n_2 = 30

9.6.4 The Sampling Distribution of the χ^2 Statistic

In these exercises, the value of Statistic must be Chi-square.

12 Run the lab for Sample from Normal(0,1) for each of the following sample sizes, and print the graphs. What is happening to the shape of the distribution as the sample size increases? Relate this to the increase in degrees of freedom. (Note: If you are unfamiliar with the χ_ν^2 distribution, you may want to refer to the *Chi-square Curves* Lab in Chapter 8.)

(a) n_1 = 3 (b) n_1 = 5 (c) n_1 = 10
(d) n_1 = 25 (e) n_1 = 30 (f) n_1 = 50

13 Let Sample from be Uniform(0,1). For each of the following values of n_1, run the lab and print the graphs. Comment on the closeness of the two distributions in each graph. Do you reach a sample size that is large enough that using the χ_{n-1}^2 distribution is a good approximation for the sampling distribution of the χ^2 statistic? How does your conclusion compare to similar circumstances for the Z and t statistics?

(a) n_1 = 3 (b) n_1 = 9 (c) n_1 = 15 (d) n_1 = 25
(e) n_1 = 30 (f) n_1 = 35 (g) n_1 = 40

14 Let Sample from be Exponential(1). For each of the following values of n_1, run the lab and print the graphs. Comment on the closeness of the two distributions in each graph. Do you reach a sample size that is large enough that using the χ_{n-1}^2 distribution is a good approximation for the sampling distribution of the χ^2 statistic? How does your result compare to the previous exercise when Sample from was Uniform(0,1)? How does your conclusion compare to similar circumstances for the Z and t statistics?

(a) n_1 = 3 (b) n_1 = 9 (c) n_1 = 15 (d) n_1 = 25
(e) n_1 = 30 (f) n_1 = 35 (g) n_1 = 40

9.6.5 The Sampling Distribution of the F Statistic

In these exercises, the value of Statistic must be F.

15 Run the lab for Sample from Normal(0,1) for each of the sample sizes given below, and print each graphs. What is happening to the shape of the distribution as the sample size increases? Relate this to the increase in degrees of freedom. Is there another sampling distribution you could use for very large n_2?

(a) n_1 = n_2 = 3 (b) n_1 = n_2 = 5 (c) n_1 = n_2 = 10
(d) n_1 = n_2 = 25 (e) n_1 = n_2 = 30 (f) n_1 = n_2 = 50

16 Let Sample from be Normal(0,1) and n_2 = 100. Run the lab for Statistic being each of Chi-square and F. Print both graphs. How do the distributions compare graphically and numerically?

17 Let Sample from be Uniform(0,1). For each of the following values of n_1 and n_2, run the lab and print the graphs. Comment on the closeness of the two distributions in each graph. Do you reach a sample size that is large enough that using the F_{n_1-1,n_2-1} distribution is a good approximation for the sampling distribution of the F statistic?

(a) n_1 = n_2 = 12 (b) n_1 = n_2 = 15 (c) n_1 = n_2 = 25
(d) n_1 = n_2 = 30 (e) n_1 = n_2 = 35 (f) n_1 = n_2 = 40

18 Let Sample from be Exponential(1). For each of the following values of n_1 and n_2, run the lab and print the graphs. Comment on the closeness of the two distributions in each graph. Do you reach a sample size that is large enough that using the F_{n_1-1,n_2-1} distribution is a good approximation for the sampling distribution of the F statistic?

(a) n_1 = n_2 = 12 (b) n_1 = n_2 = 15 (c) n_1 = n_2 = 25
(d) n_1 = n_2 = 30 (e) n_1 = n_2 = 35 (f) n_1 = n_2 = 40

10

Minimum Variance Estimation

The mean and the median are two estimators of the center of a distribution. In this lab, we will determine which of the mean or median is the best estimator for the center of the distribution for four different distributions.

10.1 Introduction

When estimating a parameter, we want the estimator we choose to be a good one. But how is "good" defined? One desirable property is that the estimator be unbiased. In this lab, we take the definition of a "good" estimator one step further. Specifically, if there is more than one unbiased estimator for a parameter, we insist that the best estimator is the unbiased estimator whose sampling distribution has the smallest, or minimum, variance among all unbiased estimators.

10.1.1 Some Basic Ideas

There are many ways to estimate the center of the distribution of a population. Two commonly used estimators are the sample mean $\bar{X}$ and the sample median $\tilde{X}$. In this lab, we are going to study which of the mean or median is the best estimator of the center of the population distribution for four population distributions, each having its center located at zero.

Two criteria are used to define exactly what it means for an estimator to be the "best estimator" of a population parameter. The first of these criteria is called *unbiasedness*. An estimator is unbiased if the mean of its sampling distribution is the parameter it is intended to estimate. For example, in the *Random Sampling* and *Central Limit Theorem* Labs, we saw that the center of the sampling distribution of $\bar{X}$ was the population mean μ. The center of the sampling distribution of $\bar{X}$ is the *mean of $\bar{X}$*. That is, for all possible samples that could be selected from the population, if we averaged the $\bar{X}$'s from all these samples, the result would be exactly the population mean μ, which is precisely what we want to estimate. Thus, the sample mean $\bar{X}$ is an unbiased estimator of the population mean μ. For the four population distributions considered in this lab, the sample median $\tilde{X}$ is also an unbiased estimator of the population mean.

The second criterion is a property known as *minimum variance*. If we have more than one unbiased estimator for a population parameter, the estimator whose sampling distribution has the smallest variance is called the *unbiased minimum variance estimator*. For example, in the *Central Limit Theorem* Lab, the variance of the sampling distribution of $\bar{X}$, also called *the variance of $\bar{X}$*, was given by σ^2/n. This means that if we were to calculate the variance of the $\bar{X}$'s we could get from every possible sample, the result would be exactly σ^2/n. If we can show that σ^2/n is smaller than the variance of all other unbiased estimators for the population mean, then $\bar{X}$ is the unbiased minimum variance estimator of μ. Minimum variance estimators estimate a parameter with smaller variability, or greater consistency, than any other estimator.

A common intuitive interpretation of these properties involves a target where the center, or "bull's-eye," is the population parameter (see Figure 10.1). The "shots" at

FIGURE **10.1**
The points in the top plot are tightly clustered (small variance) but around the wrong location (thus biased). The middle plot shows unbiasedness (equal to the bull's-eye on the average) but large variance. The bottom plot shows both unbiasedness and minimum variance.

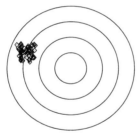

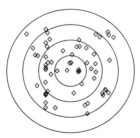

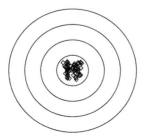

the target are the values of the estimator from different samples from the population. An unbiased estimator has shots that are evenly scattered and centered around the bull's-eye but are not necessarily tightly clustered. A minimum variance estimator has shots that are more tightly clustered than any other estimator but are not necessarily centered around the bull's-eye. Thus, an unbiased minimum variance estimator is the estimator with points that have the tightest clustering centered around the bull's-eye.

We will use two graphical measures of a distribution to compare the means and variances of the sampling distributions of the mean and median: the box plot and the histogram. The box plot is a graphical representation of the five-number summary of a data set. In this case, our data set is the 500 sample means (or medians). The bottom side of the box is the 25th percentile, or first quartile Q_1, of the 500 means (or medians). Similarly, the top side of the box is the 75th percentile, or third quartile Q_3. The line through the approximate center of the box is the median (50th percentile, second quartile) of the 500 means or medians. Note that 50% of the measurements used to construct the box plot lie between Q_1 and Q_3, within the limits of the box. The line from the bottom of the box extends to the larger of the minimum value and the *lower adjacent value* = $Q_1 - 1.5(Q_3 - Q_1)$. The line from the top of the box extends to the smaller of the maximum value and the *upper adjacent value* = $Q_3 + 1.5(Q_3 - Q_1)$. These two vertical lines are often called *whiskers*. Any small circles appearing past these lines are values that are more extreme than the lower and upper adjacent values. For more information, consult your text or the StataQuest help files.

10.2 Objectives

1. Understand the concept of minimum variance.
2. Recognize that the mean is not necessarily the best estimator of the center of a distribution.
3. Identify which of the mean or median is the best estimator for the center of the four distributions presented in the lab.

10.3 Description

To start the lab, choose `Minimum Variance Estimation` from the `Labs` menu. When the lab begins, a graphics window opens containing a graph like the one shown in Figure 10.2. The curves in this graph are the distribution functions of four populations, each having mean $\mu = 0$:

1. The Uniform(−0.5,0.5) distribution
2. The Normal(0,1) distribution
3. The Laplace(0,1) distribution
4. The t_3 distribution

If you are not familiar with these distributions, you may wish to review the *How Are Populations Distributed?* Lab.

FIGURE 10.2
Four population distribution curves: the Uniform(−0.5,0.5), Normal(0,1), Laplace(0,1), and t_3.

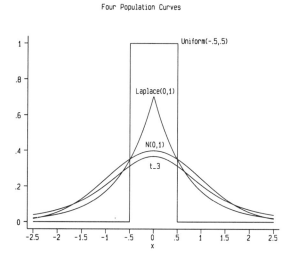

Next, a dialog box like the one shown in Figure 10.3 opens, giving a brief description of the lab. Below the description are two boxes requiring information.

1. n: The value entered in the box to the immediate right of n indicates the size of the samples to be chosen from the population distribution. The value of n must be an integer between 5 and 50. The initial value of n is 5.

2. Sample from: The distribution specified in the box to the immediate right of Sample from determines the distribution of the population from which samples are chosen. The choices for Sample from are restricted to those shown. The initial population distribution is the standard normal, or Normal(0,1), distribution.

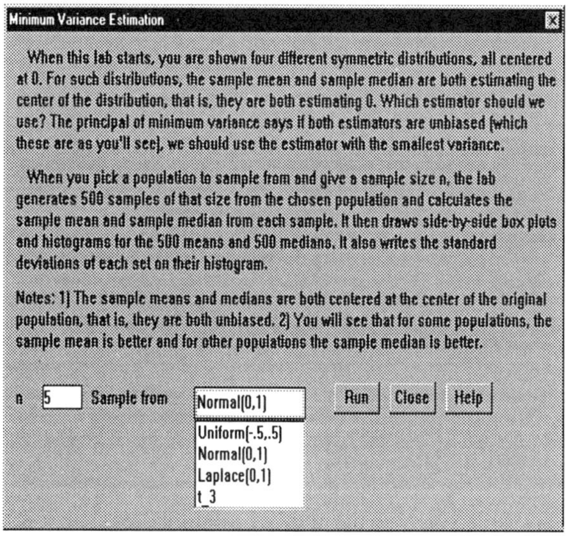

FIGURE 10.3
Dialog box for the *Minimum Variance Estimation* Lab.

To the immediate right of these boxes are three buttons.

1. **Run**: Clicking on this button causes the lab to draw 500 samples of size n from a population having the distribution specified for `Sample from`. For each sample, the sample mean $\bar{X}$ and sample median $\tilde{X}$ are calculated. While samples are being drawn and calculations are being performed, a yellow status bar will appear and slowly fill as these tasks are completed. Once they are completed, a graphics window similar to the one shown in Figure 10.4 will open. Your figure will vary slightly because the lab generates a unique set of 500 samples each time it is run. The graphics window has three sets of axes. The axes on the left side of the window contain two box plots on the same scale to make comparing them easy. The box plot on the left, drawn in yellow, is a plot of the 500 sample means. The box plot on the right, drawn in red, is a plot of the 500 sample medians. In the axes in the upper right-hand corner, also drawn in yellow, is a histogram of the 500 sample means. The red histogram immediately below that is a histogram of the 500 sample medians. The vertical axes of the two histograms are on the same scale, and the horizontal axes are aligned, so that comparing the two histograms is easier. Comparisons of the box plots and histograms will help you decide which of the sample mean or sample median is a better estimator of the population mean.

2. **Close**: Clicking on this button closes the lab and returns you to the StatConcepts menus.

3. **Help**: Clicking on this button opens a help window containing information about the lab.

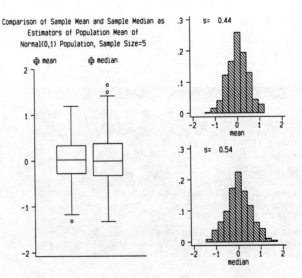

FIGURE 10.4 Comparison of the sample mean and sample median as an estimator of the population mean for samples of size $n = 5$ from a Normal(0,1) distribution.

10.4 Guided Tour of the Lab

There are four stops in the tour of minimum variance estimation. Each stop is a different population that has a symmetrical distribution centered about its mean, and all the populations have the same mean $\mu = 0$. Each stop has two major attractions: the variance of the sampling distribution of the mean and of the median. Even though each stop sounds as if it will be similar to the other stops, we think you'll be surprised at the interesting twists and turns in the tour.

Before we proceed with the tour, we need to describe a few features of the lab. In this lab, a population distribution is specified, and 500 samples of size n are selected. For each of these samples, the sample mean $\bar{X}$ and sample median $\tilde{X}$ are calculated. Then, for each of $\bar{X}$ and $\tilde{X}$, box plots and histograms of the 500 values are drawn so they may be easily compared. By studying the values of $\bar{X}$ and $\tilde{X}$ in this manner, we know how they behave for different population distributions. In practice, we get only one chance to estimate the population mean μ. By studying this lab, you will be able to decide which of the sample mean or sample median will result in the most accurate, precise estimator of μ for different population distributions.

10.4.1 The Normal(0,1) Population Distribution

For the standard normal distribution, the population mean is $\mu = 0$ and the population variance is $\sigma^2 = 1$. Which of the mean or median is the best estimator of the mean ($\mu = 0$)? To make this decision, let n = 5 and Sample from be Normal(0,1). Running the lab will open a graphics window similar to the one shown in Figure 10.4. Your figure will vary slightly because the lab generates a unique set of 500 samples each time it is run. The value s = 0.44 is the standard deviation of the 500 means. The value of s should differ slightly each time the lab is run but should be around $\sigma/\sqrt{n} = 1/\sqrt{5} = 0.4472$. The value s = 0.54 is the standard deviation of the 500 sample medians. Again, the value will vary each time the lab is run.

To determine which of the mean or median is a better estimator of the population mean, first let's compare the box plots. There are several characteristics to observe.

1 Notice the lines in the approximate centers of the boxes. In the yellow box plot, this line represents the median of the 500 means. In the red box plot, it represents the median of the 500 medians. Both lines are very close to zero. Therefore, about 50% of the 500 means (and medians) are below zero, and about 50% are above zero. These distributions are centered around zero. (Recall that the population mean is $\mu = 0$.)

2 If we compare the two boxes, the yellow box of the means is narrower (from Q_1 to Q_3) than the red box of the medians. This implies that it takes a smaller range to encompass the middle 50% of the data for the means than for the medians.

3 Finally, the whiskers of the box plot of the means do not extend as far as the whiskers for the box plots of the medians. The whiskers of the means extend from -1.5 to 1.5 while the whiskers of the medians extend from approximately -1.8 to 1.8.

These last two characteristics imply that the variability in the means is smaller than the variability in the medians.

Now compare the histograms of the means and medians. Both are symmetrical and centered at zero. But the histogram of the means is taller and narrower than the histogram of the medians. While the means range between approximately -1.5

to 1.5, the medians fall between -1.8 and 1.8 (we saw this same result in the box plots). Recall that narrower, taller distributions have smaller variances. Numerically, the standard deviation of the means is $s = 0.44$ and the standard deviation of the medians is $s = 0.54$.

In our examination of the box plots and the histograms, we saw that the sampling distributions of the mean and median were centered at the mean; that is, both the mean and median are unbiased. We also observed that the sampling distribution of the mean has a smaller variance than the sampling distribution of the median. Therefore, although both are unbiased, the sample mean is the better estimator of the population mean. In summary, based on a sample of size $n = 5$, if the population has a normal distribution, its mean is the unbiased minimum variance estimator of the sample mean.

Will changing the sample size affect these results? Let n = 20 and run the lab to find out. The result is shown in Figure 10.5. If we make comparisons similar to those done previously, we see that both the mean and median are unbiased. Both the box plots are centered around zero, and so are both of the histograms. We also see that the mean still has a smaller variance than the median, for several reasons:

1   The range of the box in the box plot of the means is narrower than for the medians.
2   The range of the means is smaller than the range of the medians.
3   The histogram of the means is taller and narrower than the histogram of the medians.
4   The standard deviation of the means s = 0.22 is smaller than the standard deviation of the medians s = 0.26.

Therefore, we can conclude that increasing the sample size does not change the fact that the mean is the best estimator of the center of a normal population.

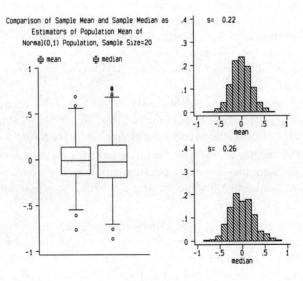

FIGURE 10.5 Comparison of the sample mean and sample median as an estimator of the population mean for samples of size $n = 20$ from a Normal(0,1) distribution.

10.4.2 The Uniform(−0.5,0.5) Population Distribution

Earlier, we suggested that the distribution would determine which of the mean or median was the better estimator. Will the conclusion reached for the normal distribution hold for the Uniform(−0.5,0.5) distribution? Change `Sample from` to `Uniform(-.5,.5)` and n to 5, and run the lab. As Figure 10.6 shows, the differences between the sampling distributions of the mean and median are striking. Everything we observed in the normal distribution case is even more marked. The first property we examined was the center of the box plots, which appear to be centered almost exactly at zero (the mean for the Uniform(−0.5,0.5) distribution is also $\mu = 0$). So do the histograms. Therefore, the sample mean and sample median both appear to be unbiased estimators of the population mean.

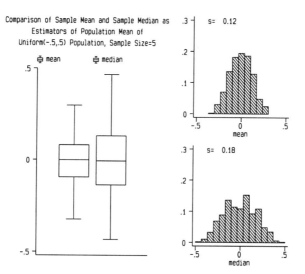

FIGURE 10.6 Comparison of the sample mean and sample median as an estimator of the population mean for samples of size $n = 5$ from a Uniform(−0.5,0.5) distribution.

When we compare the dispersion of the sampling distributions, it quickly becomes apparent that the sample mean has a smaller variance than the sample median. Graphically, the ranges of the box plots and histograms of the sample means are narrower than those of the sample medians. The histogram of the sample means is obviously taller than the histogram of the medians. Numerically, the standard deviation of the means is s = 0.12 while the standard deviation of the medians is the larger s = 0.16. We can conclude that the sample mean is by far the better estimator of the population mean. In the chapter exercises, you will examine the effect of changing sample size on this conclusion.

10.4.3 The Laplace(0,1) Population Distribution

The next distribution we will work with is the Laplace(0,1) distribution, which also has mean $\mu = 0$. If you are unfamiliar with the Laplace family of distributions, you might want to review the *How Are Populations Distributed?* Lab. Change Sample from to Laplace(0,1) and n to 5, and run the lab. The result is shown in Figure 10.7. The centers of the box plots and the histograms are still at zero, so this helps confirm the unbiasedness of both the mean and the median. But when we compare variances, the results are surprisingly contrary to what we have observed thus far. In particular, the width of the box for the box plot for the mean is now wider than for the median, as is the range of the values themselves. Further, the histogram of the medians is now the taller distribution, and s = 0.56 for the medians while for the mean s = 0.62. In this case, the median is the unbiased minimum variance estimator of the population mean.

Will increasing the sample size change this conclusion? Certainly, the unbiasedness will not be affected, but what about the variances? Change n to 20 and run the lab. When we compare the widths of the boxes of the new box plots in Figure 10.8, we see again that the median, not the mean, has the narrower box. Similarly, the range of the median is smaller than the range of the mean. The medians' histogram is taller and narrower. And s = 0.25 for the median, which is less than s = 0.32 for the means. Thus, changing the sample size did not change our conclusion that the median is the best estimator of the population mean for a Laplace distribution.

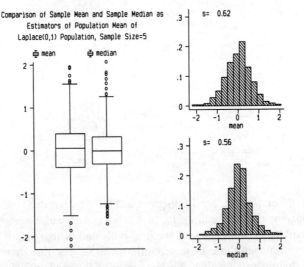

FIGURE 10.7 Comparison of the sample mean and sample median as an estimator of the population mean for samples of size $n = 5$ from a Laplace(0,1) distribution.

10.4.4 The t_3 Population Distribution

The final population distribution we will compare these two estimators for is the t_3 distribution. We saw in the *t Converging to Z* Lab that as the degrees of freedom increase, the t_ν distribution becomes more and more like the Z distribution. Based on this knowledge, our first intuition is that the best estimator for the mean $\mu = 0$ of the t_3 distribution is the mean.

Change n to 5 and Sample from to t_3, and run the lab. Again, we see in Figure 10.9 that the distributions of both estimators are centered around zero, so that the unbiasedness is confirmed. This is no surprise. But what is surprising is that the range of the box of the median is smaller than the range of the box of the mean. The range of

FIGURE 10.8 Comparison of the sample mean and sample median as an estimator of the population mean for samples of size $n = 20$ from a Laplace(0,1) distribution.

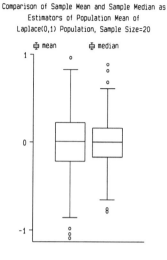
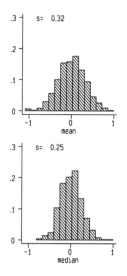

FIGURE 10.9 Comparison of the sample mean and sample median as an estimator of the population mean for samples of size $n = 5$ from a t_3 distribution.

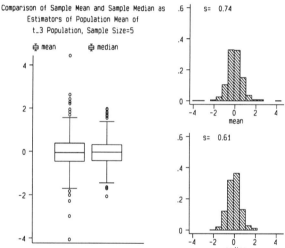

the median is smaller than the range of the mean. And the histogram of the median is narrower and taller than the histogram of the mean! To cap it off, $s = 0.61$ for the median and $s = 0.74$ for the mean. In the case of the t_3 distribution, the median, not the mean, is the unbiased minimum variance estimator of μ!

10.5 Summary

Although this lab was shorter than some, many properties and concepts were introduced.

1. If the mean of the sampling distribution of an estimator is the parameter the estimator is intended to estimate, then the estimator is *unbiased*.

2. If there is more than one unbiased estimator for a parameter, then the estimator with *minimum variance* is the estimator whose sampling distribution has smaller variance than all other unbiased estimators.
3. The *best estimator* of a population parameter is the unbiased minimum variance estimator.
4. For a normally distributed population, the mean is the best estimator for the population mean.
5. For a uniformly distributed population, the mean is the best estimator for the population mean.
6. For a population having a Laplace distribution, the median is the best estimator for the population mean.
7. For a population having a t_3 distribution, the median is the best estimator for the population mean.
8. The sample size does not affect which estimator is the best estimator.

10.6 Lab Exercises

1. What is meant by an estimator being unbiased? Why is this a desirable property?
2. What is meant by an unbiased estimator having minimum variance? Why is this a desirable property?
3. What is meant by the best unbiased estimator?

10.6.1 The Normal(0,1) Population Distribution

4. Let Sample from be Normal(0,1). Run the lab for the following values of n, and print each of the four resulting graphics windows. How do the box plots and the histograms compare? Does it appear that both the mean and median are unbiased estimators of the population mean? For each sample size, which of the mean or median is the better estimator of the population mean? How does increasing the sample size affect the sampling distributions of the two estimators, and which of the two estimators is the best estimator?

   (a) n = 10   (b) n = 30   (c) n = 40   (d) n = 50

10.6.2 The Uniform(−0.5, 0.5) Population Distribution

5. Let Sample from be Uniform(−.5, .5). Run the lab for the following values of n, and print each of the five resulting graphics windows. How do the box plots and the histograms compare? Does it appear that both the mean and median are unbiased estimators of the population mean? For each sample size, which of the mean or median is the better estimator of the population mean? How does increasing the sample size affect the sampling distributions of the two estimators, and which of the two estimators is the best estimator?

   (a) n = 10   (b) n = 20   (c) n = 30   (d) n = 40   (e) n = 50

10.6.3 The Laplace(0,1) Population Distribution

6 Let Sample from be Laplace(0,1). Run the lab for the following values of n, and print each of the four resulting graphics windows. How do the box plots and the histograms compare? Does it appear that both the mean and median are unbiased estimators of the population mean? For each sample size, which of the mean or median is the better estimator of the population mean? How does increasing the sample size affect the sampling distributions of the two estimators, and which of the two estimators is the best estimator?

(a) n = 10   (b) n = 30   (c) n = 40   (d) n = 50

10.6.4 The t_3 Population Distribution

7 Let Sample from be t_3. Run the lab for the following values of n, and print each of the five resulting graphics windows. How do the box plots and the histograms compare? Does it appear that both the mean and median are unbiased estimators of the population mean? For each sample size, which of the mean or median is the better estimator of the population mean? How does increasing the sample size affect the sampling distributions of the two estimators, and which of the two estimators is the best estimator?

(a) n = 10   (b) n = 20   (c) n = 30   (d) n = 40   (e) n = 50

11

Interpreting Confidence Intervals

A *confidence interval* for a parameter is an interval on the number line that we can be confident contains the value of that parameter. For example, if we find that the interval (90,110) is a 95% confidence interval for the mean μ of the population of IQs in the United States, then we can be 95% confident that the average of the population of IQs lies somewhere between 90 and 110.

11.1 Introduction

This lab studies confidence intervals for six cases, which may be classified as follows:

1. Continuous populations
   (a) One-sample procedures
      i. *Case 1*: The population mean μ
      ii. *Case 2*: The population variance σ^2
   (b) Two-sample procedures
      i. *Case 3*: The difference between two population means $\mu_1 - \mu_2$ (Note: If two populations have the same mean, the difference of the means is 0.)
      ii. *Case 4*: The ratio of two population variances σ_1^2/σ_2^2 (Note: If two populations have the same variance, the ratio of the variances is 1.)

2 Discrete (or 0–1) populations
   (a) *Case 5*: The one-sample procedure for the population proportion π
   (b) *Case 6*: The two-sample procedure for the difference between two population proportions $\pi_1 - \pi_2$ (Note: If two populations have the same proportion of 1's, the difference of the proportions is 0.)

The six cases in this classification system correspond to the six choices under `Parameter` in the dialog box shown in Figure 11.1, later in the chapter.

11.1.1 Some Basic Ideas

Before using the lab, there are five important ideas to understand:

1 *Meaning of confidence:* If all of the assumptions of a confidence interval procedure are satisfied (see item 5 below) and we randomly select millions of samples from a population and calculate a $100(1 - \alpha)\%$ confidence interval for a parameter of the population, then the percentage of intervals containing the true value of the parameter is, in fact, $100(1 - \alpha)$. In practice, we get only one sample and thus only one interval. But if we know that $100(1 - \alpha)\%$ of all samples have confidence intervals containing the true value of the parameter, then we can feel confident that our particular interval does.

2 *Using confidence intervals to test conjectures:* If we conjecture a particular value for one of the above parameters and the conjectured value is contained in the confidence interval for the parameter, then the conjecture is not unreasonable. For example, if we believe that two populations have the same mean values—that is, $\mu_1 = \mu_2$, and 0 is within the confidence interval for $\mu_1 - \mu_2$—then this belief is not unreasonable.

3 *Width of intervals:* For a specified level of confidence (such as 95%), we want our confidence interval to be as narrow as possible. For example, if a 95% confidence interval for mean daily summer temperature in Texas turns out to be (50,150), we really haven't learned anything because we know from experience that the mean temperature is in that interval. For it to be useful, we would need an interval that has a width of 1 or 2 degrees. We will see that it is sample size that determines widths of intervals; the bigger sample sizes are, the narrower the intervals are.

4 *Formulas for confidence intervals:* The formulas for the intervals can be derived immediately from the ideas in the *Sampling Distributions* Lab. For example, we saw in that lab that from a sample of size n,

$$t = \frac{\bar{X} - \mu}{s/\sqrt{n}}$$

has a 95% chance of lying between $t_{0.975, n-1}$ and $t_{0.025, n-1}$; that is,

$$P\left(t_{0.975, n-1} \leq \frac{\bar{X} - \mu}{s/\sqrt{n}} \leq t_{0.025, n-1}\right) = 0.95.$$

(See the *Critical Values* Lab for more about the notation $t_{0.975,n-1}$ and other, similar symbols.) If the inequality is solved to get μ by itself in the middle, we find

$$P\left(\bar{X} + t_{0.975,n-1}\frac{s}{\sqrt{n}} \leq \mu \leq \bar{X} + t_{0.025,n-1}\frac{s}{\sqrt{n}}\right) = 0.95,$$

and thus (because $t_{0.975,n-1} = -t_{0.025,n-1}$), $\bar{X} \pm t_{0.025,n-1}s/\sqrt{n}$ is a 95% confidence interval for μ. If from a random sample of $n = 25$ values from a N(0,1) population we found $\bar{X} = 0.15$ and $s = 0.9$, we would find $0.15 \pm t_{0.025,24}0.9/\sqrt{25}$ or $0.15 \pm (2.492)(0.9)/5$ or $(-0.299, 0.599)$ as the confidence interval for the mean of the population. For a general level of confidence, $(1-\alpha)$, we would get $\bar{X} \pm t_{\alpha/2,n-1}s/\sqrt{n}$ as a $100(1-\alpha)\%$ confidence interval for μ.

The formulas for the confidence intervals calculated in the lab are as follows:

(a) *Case 1:* For the mean μ of a population given a random sample of size n from that population:

$$\bar{X} \pm t_{\alpha/2,n-1}\frac{s}{\sqrt{n}}.$$

(b) *Case 2:* For the variance σ^2 of a population given a random sample of size n from that population:

$$\left(\frac{(n-1)s^2}{\chi^2_{\alpha/2,n-1}}, \frac{(n-1)s^2}{\chi^2_{1-\alpha/2,n-1}}\right).$$

(c) *Case 3:* For the difference $\mu_1 - \mu_2$ between the means of two populations given independent samples of size n_1 and n_2 from those populations:

$$(\bar{X}_1 - \bar{X}_2) \pm t_{\alpha/2,n_1+n_2-2}\sqrt{s_P^2\left(\frac{1}{n_1} + \frac{1}{n_2}\right)},$$

where the pooled variance estimator s_P^2 is given by

$$s_P^2 = \frac{(n_1-1)s_1^2 + (n_2-1)s_2^2}{n_1 + n_2 - 2}.$$

(d) *Case 4:* For the ratio σ_1^2/σ_2^2 of variances of two populations given independent samples of size n_1 and n_2 from those populations:

$$\left(\frac{s_1^2/s_2^2}{F_{\alpha/2,n_1-1,n_2-1}}, \frac{s_1^2/s_2^2}{F_{1-\alpha/2,n_2-1,n_1-1}}\right).$$

(e) *Case 5:* For the proportion π of 1's in a 0–1 population given a random sample of size n from the population:

$$p \pm Z_{\alpha/2}\sqrt{\frac{p(1-p)}{n}},$$

where p is the proportion of 1's in the sample.

(f) *Case 6:* For the difference $\pi_1 - \pi_2$ between proportions of 1's in two 0–1 populations given independent random samples of size n_1 and n_2 from those populations:

$$(p_1 - p_2) \pm Z_{\alpha/2} \sqrt{\frac{p_1(1-p_1)}{n_1} + \frac{p_2(1-p_2)}{n_2}},$$

where p_1 and p_2 are the proportion of 1's in the two samples.

5 *Assumptions being made:* For the confidence level (such as 95%) to be valid, certain assumptions about the sampling method and the populations being sampled from must be satisfied:

(a) In all cases, the samples being obtained must be *random samples* (see the *Random Sampling* Lab).

(b) The formulas for cases 1–4 are valid under two circumstances:

   i For any sample size when the distribution of the population from which the sample is being taken belongs to the family of normal distributions.

   ii For a continuous population with any distribution when the sample size(s) is "large."

(c) In cases 3, 4, and 6, the samples from the two populations being sampled from must be *independent*. This rules out, for example, any paired data such as "before and after" studies.

(d) The formulas in cases 5 and 6 assume that the smaller of the *expected number of 1's or 0's* in any sample must be at least 5. The expected number of 1's for a sample of size n from a 0–1 population that has proportion π of 1's is $n\pi$, while the expected number of 0's is $n(1-\pi)$. See the *Normal Approximation to Binomial Probabilities* Lab for more information on this issue.

11.2 Objectives

1 Understand what is meant by *confidence* and why we can be so confident.
2 Illustrate how intervals get wider (narrower) as confidence increases (decreases).
3 Illustrate how intervals get narrower (wider) as sample sizes increase (decrease).
4 Examine the effect of violation of assumptions.

11.3 Description

To start the lab, choose Interpreting Confidence Intervals from the Labs menu. The dialog box shown in Figure 11.1 will open. This dialog box contains a host of information, including a brief narrative on how to use the lab. Below that narrative are six boxes in which you will supply information.

1 n_1: To the immediate right of n_1 is a box where the value of a sample size is to be entered. The initial value of n_1 is 20. If you are using a one-sample

FIGURE 11.1
Dialog box for the *Interpreting Confidence Intervals* Lab.

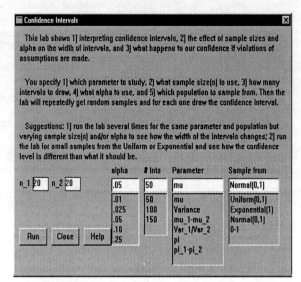

procedure, type the value of the sample size here. If you are using a two-sample procedure, type the value for the size of the sample from the first population. The values of n_1 are restricted to numbers between 5 and 500.

2. n_2: To the immediate right of n_2 is a box where you enter the value. The initial value of n_2 is 20. If you are using a one-sample procedure, this value is ignored. If you are using a two-sample procedure, type in the value for the size of the sample from the second population. The values of n_2 are restricted to numbers between 5 and 500.

3. alpha: Directly beneath the label alpha are values that may be chosen so that the level of confidence for the intervals calculated is 1− alpha. For example, the initial value is 0.05, so the initial level of confidence for the intervals calculated is 95%. The value of alpha is not restricted to those shown. To calculate confidence intervals for a value of alpha not supplied, simply type in the desired value.

4. # Ints: Directly beneath this label are values that indicate the number of samples to be drawn. The initial value is 50. For each sample drawn, a $100(1 - \text{alpha})$ confidence interval is calculated. The value of # Ints is not restricted to those shown. To have a number of samples drawn for a value that is not supplied, simply type in the desired value for # Ints.

5. Parameter: Below this label are six different cases that correspond to the six confidence intervals outlined under item 4 in Subsection 11.1.1. The initial value is mu. This will calculate confidence intervals for the population mean μ. To calculate confidence intervals for the ratio of two variances σ_1^2/σ_2^2, choose Var_1/Var_2. The values for Parameter are restricted to those shown.

6. Sample from: Below this label are the names of four different distributions. The distribution chosen specifies the distribution of the population from which the samples are being taken. The initial distribution is the Normal(0,1) distribution. The distributions for Sample from are restricted to those shown:

(a) A Uniform(0,1) distribution with mean $\mu = 0.50$ and variance $\sigma^2 = 1/12$.
(b) An Exponential(1) distribution with mean $\mu = 1$ and variance $\sigma^2 = 1$.
(c) A Normal(0,1) distribution with mean $\mu = 0$ and variance $\sigma^2 = 1$.
(d) A 0–1 population with proportion of 1's $\pi = 0.50$ and variance $\pi(1 - \pi) = 0.25$.

In the bottom left-hand corner of the dialog box are three buttons.

1. Run: Before you click on Run, be sure a value has been specified for all of the boxes just described. When you click on this button, a graphics window opens (see Figure 11.2). The caption includes information on the confidence level, the population from which samples are being drawn, and the sample size(s). A red vertical line representing the value of the population parameter appears on the graph. The computer randomly generates # Ints sets of n numbers from the specified population in Sample from and for each set of numbers calculates the $100(1 - \text{alpha})$ confidence interval for the parameter specified in the Parameter box. As each interval is calculated, a horizontal line beginning at the lower confidence limit and ending at the upper confidence limit is drawn. If the confidence interval contains the value of the parameter, the horizontal line will intersect the vertical line and will be drawn in red. If the confidence interval does not contain the value of the parameter, the horizontal line will not intersect the vertical line and will be drawn in yellow. After all # Ints samples are chosen and all confidence intervals are calculated, a second heading appears at the top of the graphics window giving the percentage of intervals that contained the value of the parameter for this run of the lab. You may click on Run as many times as you wish.

2. Close: Clicking on this button closes the lab and returns you to the StatConcepts menus.

3. Help: Clicking on this button opens a help window containing information about the lab.

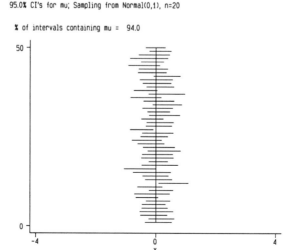

FIGURE 11.2
Fifty 95% confidence intervals for the mean μ of a N(0,1) population with $n = 20$ for each interval.

11.4 Guided Tour of the Lab

We start our tour with the simplest case, namely calculating confidence intervals for the mean of a population whose distribution belongs to the family of normal distributions. Note that, when the lab first begins, the mean mu is already specified in the Parameter box. We will use the initial values in all the boxes. Thus we will be calculating fifty 95% confidence intervals using samples of size 20 from a Normal(0,1) population. Finally, since the population we are sampling from is Normal(0,1), the mean of that population is $\mu = 0$. This is represented by a red vertical line above 0.

Before we proceed, one final note is in order. In practice, we do not know the value of the mean (the parameter) of the population, and a single confidence interval is used to estimate the population mean (or parameter) with a specific degree of "confidence." One objective of this lab is to examine the behavior of all confidence intervals. To do this, many random samples are generated, and a confidence interval for each sample is calculated. This allows us to observe their performance in a controlled situation. By doing so, we will learn what to expect when we actually use these confidence intervals and will develop an understanding of what is truly meant by "95% confident."

To begin, click on Run. This causes the lab to randomly select the 50 samples, each containing 20 observations. For each sample, the lab calculates the sample mean $\bar{X}$ and the sample standard deviation s, and then finds the lower and upper limits of the confidence interval using

$$\bar{X} \pm t_{0.025, 19} \frac{s}{\sqrt{20}}.$$

For each sample, a horizontal line is drawn above the horizontal axis from the lower confidence limit to the upper confidence limit.

One possible result was shown in Figure 11.2. You will probably get a slightly different graph because the lab selects a different set of 50 samples every time it is run. In this figure, 47 of the 50 samples (or 94%) have confidence intervals containing the true mean value of 0. Can you find the three intervals in the figure that don't "capture" the true mean? When you run the lab, how many of your 50 samples have intervals containing 0?

You can rerun the lab under these same conditions by clicking on Run. Each time you do, 50 new samples are selected, and the corresponding confidence intervals are calculated and drawn. For each set of intervals, you should notice that the percentage containing the population mean will be close to 95%. You should also pay close attention to the width of the intervals.

11.4.1 The Effect of Sample Size

To see the effect of sample size on the width of a confidence interval, change the sample size in n_1 from 20 to 5 and click on Run. The confidence intervals calculated are now of the form

$$\bar{X} \pm t_{0.025, 4} \frac{s}{\sqrt{4}}.$$

The result is shown in Figure 11.3 (again, your result should be similar but probably will not be the same). Notice that this figure is on the same scale (−4 to 4 on the horizontal axis) so that we can easily compare the widths of the intervals from the two sample sizes. For the smaller sample size, the widths of the confidence intervals are larger. We would hope this is the case, because it means that having more data (a larger sample size) will let us estimate the parameter more precisely.

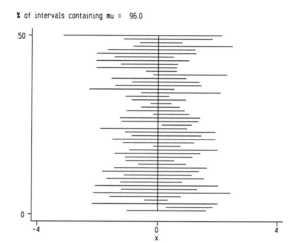

FIGURE 11.3 Fifty 95% confidence intervals for the mean μ of a $N(0,1)$ population with $n = 5$ for each interval.

Note that the probability of capturing the true value of the parameter is the same in both figures because we used $\alpha = 0.05$ in both. Stated another way, a larger sample size has no effect on the probability of capturing the value of the parameter in the confidence interval! The sample size affects only the precision of our estimator.

11.4.2 The Effect of Confidence Level

To see the effect of α on the behavior of confidence intervals, change n_1 back to 20, select alpha = 0.25, and run the lab again. The confidence intervals are now of the form

$$\bar{X} \pm t_{0.125,19} \frac{s}{\sqrt{20}}.$$

The result is shown in Figure 11.4. Notice that the intervals are narrower than in the $n = 20$ and $\alpha = 0.05$ case, but a much larger percentage of intervals do not contain the population mean. This is what being only 75% confident means: The chance that any one sample will result in a confidence interval containing the population mean is now only 75%. In other words, now that $\alpha = 0.25$, of all the possible samples we could choose, only 75% of them will result in a confidence interval containing the true value of the mean.

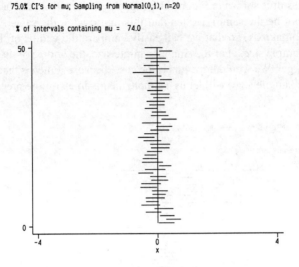

FIGURE 11.4
Fifty 75% confidence intervals for the mean μ of a N(0,1) population with $n = 20$ for each interval.

11.4.3 Violation of Assumptions

As noted previously, the confidence interval formula for μ is only valid under two conditions:

1. For any sample size when the distribution of the population from which the sample is being taken belongs to the family of normal distributions.

2. For a continuous population with any distribution when the sample size is "large."

We will now examine the effect of violating these assumptions. Change the value of n_1 to 5 and the distribution in `Sample from` to `Exponential(1)`. (For this distribution, the population mean $\mu = 1$.) Graphs of the exponential curves are given in the *How Are Populations Distributed?* Lab. The exponential distribution with $\mu = 1$ is not shown, but by observing the trend in those graphs, you can see that the Exponential(1) distribution is even more skewed than those shown in that lab. From these graphs, you can see that such populations are not bell-shaped at all, but rather are heavily skewed to the right.

The result of running the lab for $\alpha = 0.05$ is shown in Figure 11.5. Notice that instead of capturing the true mean approximately 95% of the time, the true mean was in only 42 of the 50 intervals, or 84% of the time. This shows how important such a violation of assumptions can be. Instead of being 95% confident, we can, in fact, be only around 84% confident.

11.4.4 Intervals for Other Parameters

We just concluded the part of our tour that examined the behavior of confidence intervals for a population mean. For the rest of this tour, we will examine the behavior of confidence intervals for the remaining five cases. As we continue on, notice that even though the parameters change and the appearance of the confidence intervals may change somewhat, the following still apply:

FIGURE 11.5 Fifty confidence intervals for the mean μ of an Exponential(1) population.

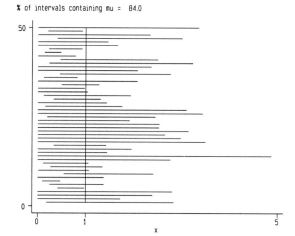

1. If the assumptions are satisfied, then to be "95% confident" (for example) means that out of all possible samples of size n from the population, 95% of the samples will result in a confidence interval containing the true value of the population parameter. Stated in another way, if we could take many, many samples of size n from the population, 95% of those samples would result in a confidence interval containing the true value of the population parameter.

2. The sample size determines the width of the confidence interval. The larger the sample size is, the narrower the width of the confidence interval. Thus, the larger a sample size we have, the more precise our confidence interval becomes.

3. If the assumptions are violated, the level of confidence is compromised.

Confidence Intervals for the Variance σ^2 of a Population

Let `n_1 = 20`, `alpha = 0.05`, and `# Ints = 50`, and choose `Parameter` of `Variance` and `Sample from` of `Normal(0,1)`. The normally distributed population we are sampling from has variance $\sigma^2 = 1$. The vertical red line drawn above 1 represents that value for the population variance.

Run the lab. The resulting graph will look something like what is shown in Figure 11.6. As before, the horizontal lines begin at the lower confidence limit and end at the upper confidence limit. If the horizontal line is red, it intersects the vertical red line, and the confidence interval it represents "captures" the value of the population variance. If the confidence interval does not capture the population variance, the horizontal line will be yellow.

In Figure 11.6, 94% of the confidence intervals contain the value of the population variance ($\sigma^2 = 1$). You should obtain a similar result. Because the population we are sampling from belongs to the family of normal distributions, 95% of all samples of size n will result in confidence intervals containing the true value of σ^2. We observed only a small subset of all possible samples, so we did not get exactly 95%, but we did come close!

FIGURE 11.6
Fifty 95% confidence intervals for the variance σ^2 of a N(0,1) population with $n = 20$ for each interval.

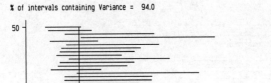
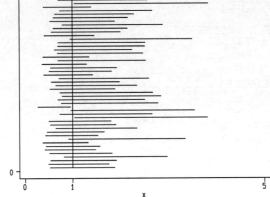

Now change the value of n_1 to 5 and run the lab again. We can see in Figure 11.7 that the confidence intervals are noticeably wider. The precision of the confidence interval has suffered greatly! But the probability of capturing the value $\sigma^2 = 1$ has not changed.

FIGURE 11.7
Fifty 95% confidence intervals for the variance σ^2 of a N(0,1) population with $n = 5$ for each interval.

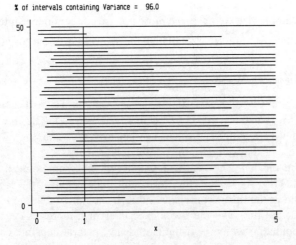

Finally, to observe the effects of violating the normality assumption, change the distribution in `Sample from` to `Exponential(1)`. For this population, the variance is $\sigma = 1$. We can see once again that if the assumptions are not satisfied, our definition of confidence is compromised. In Figure 11.8, only 82% of the confidence intervals contain the true value of the variance of this new population.

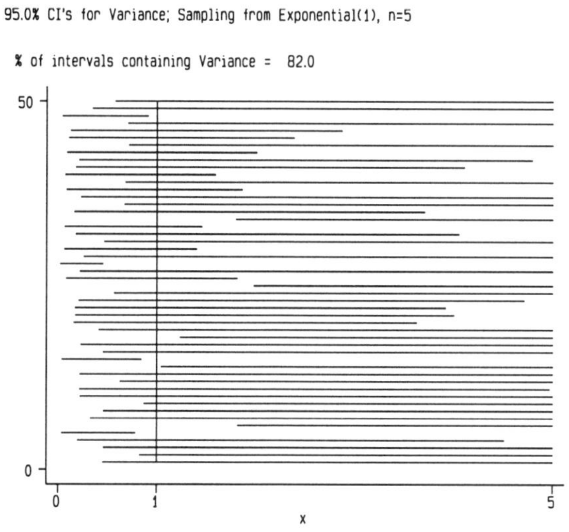

FIGURE 11.8
Fifty confidence intervals for the variance σ^2 of an Exponential(1) population.

95.0% CI's for Variance; Sampling from Exponential(1), n=5

% of intervals containing Variance = 82.0

Confidence Intervals for the Difference Between Two Population Means $\mu_1 - \mu_2$

Confidence intervals for the difference between two parameters are often used to compare the values of the parameters or to decide if the two parameters could be equal. For example, if the confidence interval for the difference $\mu_1 - \mu_2$ in two population means contains 0, then 0 is a plausible value for $\mu_1 - \mu_2$, and so we would say that the two means *could be* equal. Notice we do not say that the means *are* equal, because their difference could also be any other value contained in the confidence interval. However, because 0 is one of those values, we cannot rule it out, nor can we rule out that the means *could be* equal. On the other hand, if 0 is not contained in the confidence interval, then we are $100(1-\alpha)\%$ certain that $\mu_1 - \mu_2$ could not be 0, and hence the two means could not be equal.

When you choose mu_1-mu_2 in the Parameter box, the samples are chosen from two populations with the same mean. Therefore, $100(1-\alpha)\%$ of all confidence intervals for $\mu_1 - \mu_2$ will contain zero. Run this lab for varying combinations of sample sizes, values of α, and population distributions. You will observe the same types of behavior as you did for confidence intervals for μ and σ^2.

1 When the population distribution in Sample from is the Normal(0,1) distribution, for any sample sizes the percentage of confidence intervals containing the true value of the difference $\mu_1 - \mu_2$, which is 0, will be approximately $100(1-\alpha)$. In the long run, this percentage will be exactly $100(1-\alpha)$.

2 When the values of the sample sizes become larger, the confidence intervals become narrower. The larger our sample sizes, the more precise the confidence interval estimator becomes.

3 When the value of α gets smaller, the percentage of confidence intervals that do not contain zero also becomes smaller and the percentage of confidence intervals that do contain zero becomes larger. The confidence intervals are becoming more reliable at capturing the true value of $\mu_1 - \mu_2$.

4 If the assumptions are violated (for example, taking samples from Exponential(1) distributions with small sample sizes), the level of confidence is compromised; that is, the probability of capturing the true value of $\mu_1 - \mu_2$ is smaller than $100(1-\alpha)$.

Confidence Intervals for the Ratio of Two Population Variances σ_1^2/σ_2^2

In the same manner that the confidence interval for the difference between two population means can be used to "test" whether the two population means are equal, a confidence interval for the ratio of two population variances σ_1^2/σ_2^2 may be used to determine if the variances are equal. Because we are now considering a ratio, the variances could be equal if the confidence interval for their ratio contains one (as opposed to zero for the difference in two means).

Choose Var_1/Var_2 in the Parameter box and run the lab for varying combinations of sample sizes, values of α, and distributions for the populations in Sample from. What do you observe for similar scenarios considered for previous confidence intervals?

Confidence Intervals for the Proportion π of 1's in a 0–1 Population

The confidence intervals calculated in this lab are used for estimating the proportion π of 1's in a 0–1 population (so Sample from must be 0-1). In this lab, the true value of π is 0.5. In the *Sampling From 0–1 Populations* Lab, we saw that if the smaller of the expected number of successes ($n\pi$) and the expected number of failures ($n(1-\pi)$) was at least 5, then the normal distribution was a good approximation for the binomial distribution. Based on this idea, we can use the normal distribution to construct a confidence interval for π if the smaller of $n\pi$ and $n(1-\pi)$ is at least 5.

In the Parameter box, specify pi. Let n_1 = 20, alpha = 0.05, and # Ints = 50, and be sure Sample from is 0-1. (Note that $n\pi = (20)(0.5) = 10$ and $n(1-\pi) = 20(1-0.5) = (20)(0.5) = 10$, so that the assumption is satisfied.) Click on Run. The result should look something like what is shown in Figure 11.9. Approximately 95% of the red horizontal lines representing the confidence intervals intersect the red vertical line representing the true value of π, 0.5.

Now change the value of n_1 to 5. As expected, the confidence intervals become much wider. But something else is happening, too. Click on Run several times and watch the value of the confidence level closely. You should notice that it is usually smaller than 95%. This is because the smaller of $n\pi = (5)(0.5) = 2.5$ and $n(1-\pi) = (5)(1-0.5) = 2.5$ is not large enough. For this small sample size, our assumption has been violated. And, as with other confidence intervals, when the assumptions are violated, the level of confidence is compromised.

Run the lab for varying sample sizes and values of α. (Remember, we're sampling from a 0–1 population, so the Sample from box must remain 0-1.) For larger sample sizes, are the confidence intervals becoming more precise? For smaller values of α, are they becoming more reliable in capturing $\pi = 0.5$?

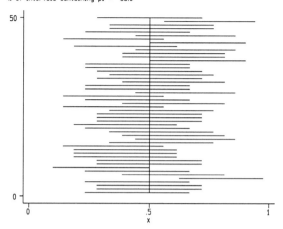

FIGURE 11.9
Fifty 95% confidence intervals for the proportion π of 1's in a 0–1 population with $n = 20$ for each interval.

Confidence Intervals for the Difference $\pi_1 - \pi_2$ in the Proportion of 1's From Two Independent 0–1 Populations

Confidence intervals for the difference $\pi_1 - \pi_2$ are used to compare the proportion of 1's in two different, independent 0–1 populations. In this lab, the proportions from the two populations are both equal to 0.5, so the confidence intervals for $\pi_1 - \pi_2$ should contain zero $100(1 - \alpha)\%$ of the time in the long run if the assumptions are satisfied. For both populations, we assume that the smaller of the expected number of successes and the expected number of failures is at least 5.

Be sure the Sample from is 0-1. Choose pi_1-pi_2 in the Parameter box, and run the lab for varying sample sizes and values of α. Observe what happens to the widths of the confidence intervals as the sample sizes change. What happens to the reliability of the confidence intervals containing zero as the value of α changes?

11.5 Summary

The phrase "$100(1 - \alpha)\%$ confident" means that in the long run, if many samples are selected and a confidence interval is calculated for each sample, then exactly $100(1 - \alpha)\%$ of those samples will yield a confidence interval containing the true value of the population parameter. As we examined the behavior of confidence intervals for many different parameters, we witnessed three major properties:

1. As the sample size(s) increases, the width of the confidence interval decreases. The confidence interval gives a more precise estimate of the value of the parameter when it contains more information (data). If the assumptions are satisfied, increasing the sample size has no effect on how reliable the confidence interval is in capturing the true value of the population parameter.

2. As the value of α decreases, the confidence intervals become more reliable in capturing the true value of the parameter. If the assumptions are satisfied, in the long run, we can expect $100(1 - \alpha)\%$ of the samples to contain the true value of

the population parameter. The decreasing value of α has very little effect on the width of the confidence interval.

3. If the assumptions are violated, the definition of $100(1-\alpha)\%$ is compromised. More specifically, the percentage of confidence intervals containing the true value of the parameter is less than $100(1-\alpha)\%$ when the assumptions are violated. The only remedy is to increase the sample size until it is "large enough." How large the sample size needs to be depends on the parameter being estimated and the distribution of the parent population. If the parent population is severely skewed, a larger sample size is necessary.

11.6 Lab Exercises

For each of the parameters $\mu, \sigma^2, \mu_1 - \mu_2$, and σ_1^2/σ_2^2, answer the first four questions. The following information will be helpful.

- For the Normal(0,1) distribution, $\mu = 0$ and $\sigma^2 = 1$.
- For the Uniform(0,1) distribution, $\mu = 0.5$ and $\sigma^2 = 1/12$.
- For the Exponential(1) distribution, $\mu = 1$ and $\sigma^2 = 1$.
- In any two-sample procedure, the parameters from the two populations will be equal. Therefore, the true value of $\mu_1 - \mu_2 = 0$ and the true value of $\sigma_1^2/\sigma_2^2 = 1$.

1. For the following combinations of # Ints and alpha, calculate confidence intervals for the population parameter from a Normal(0,1) population(s), using a sample(s) of size 20. Print each graph. For each of the combinations, how many of the confidence intervals do you expect to contain the true value of the population parameter? How many did you observe when you ran the lab?

   (a) # Ints = 100, alpha = 0.25 (b) # Ints = 100, alpha = 0.10
   (c) # Ints = 100, alpha = 0.01 (d) # Ints = 150, alpha = 0.25
   (e) # Ints = 150, alpha = 0.10 (f) # Ints = 150, alpha = 0.01

2. Let # Ints = 100. For the following sample sizes, calculate 95% confidence intervals for the population parameter from a Normal(0,1) distribution(s). Print each graph. As the sample sizes increase, what happens to the widths of the confidence intervals? Is the percentage of times the confidence intervals capture the true value of the population parameter affected by these changing sample sizes?

   (a) n_1 = 10, (n_2 = 10)   (b) n_1 = 25, (n_2 = 25)
   (c) n_1 = 50, (n_2 = 50)   (d) n_1 = 100, (n_2 = 100)

3. Let # Ints = 100. For the following combinations of parent populations and sample sizes, calculate 90% confidence intervals for the population parameter. Print each graph. As the sample sizes increase, what happens to the widths of the confidence intervals? Is the percentage of times the confidence intervals capture the true value of the population parameter affected by these changing sample sizes?

   (a) Sample from = Uniform(0,1), n_1 = 5, (n_2 = 5)
   (b) Sample from = Uniform(0,1), n_1 = 10, (n_2 = 10)

(c) `Sample from = Uniform(0,1)`, n_1 = 20, (n_2 = 20)
(d) `Sample from = Uniform(0,1)`, n_1 = 30, (n_2 = 30)
(e) `Sample from = Uniform(0,1)`, n_1 = 50, (n_2 = 50)
(f) `Sample from = Exponential(1)`, n_1 = 5, (n_2 = 5)
(g) `Sample from = Exponential(1)`, n_1 = 10, (n_2 = 10)
(h) `Sample from = Exponential(1)`, n_1 = 20, (n_2 = 20)
(i) `Sample from = Exponential(1)`, n_1 = 30, (n_2 = 30)
(j) `Sample from = Exponential(1)`, n_1 = 50, (n_2 = 50)

4 Answer the following questions using sample sizes 5, 10, 15, 20, 25, and 30.

   (a) For the `Uniform(0,1)` distribution, for each of the six sample sizes, how close to 90% is the observed percentage of intervals containing the true value of the parameter? What happens to this percentage as the sample size increases? Based on what you observed, for the `Uniform(0,1)` distribution, how large does n need to be so that we can say the confidence level is not compromised?

   (b) For the `Exponential(1)` distribution, for each of the six sample sizes, how close to 90% is the observed percentage of intervals containing the true value of the parameter? What happens to this percentage as the sample size increases? Based on what you observed, for the `Exponential(1)` distribution, how large does n need to be so that we can say the confidence level is not compromised?

The remainder of the exercises will help you study the behavior of confidence intervals calculated for population proportions when sampling from a 0–1 population. Be sure to change the choice in `Sample from` to 0-1. Do exercises 5 and 6 for both $\pi = 0.5$ and $\pi_1 - \pi_2 = 0$.

5 For the following combinations of # `Ints` and `alpha`, calculate confidence intervals for the population parameter using a sample(s) of size 20. Print each graph. For each of the combinations, how many of the confidence intervals do you expect to contain the true value of the population proportion ($\pi = 0.50$)? How many did you observe when you ran the lab?

   (a) # `Ints` = 100, `alpha` = 0.25   (b) # `Ints` = 100, `alpha` = 0.10
   (c) # `Ints` = 100, `alpha` = 0.01   (d) # `Ints` = 150, `alpha` = 0.25
   (e) # `Ints` = 150, `alpha` = 0.10   (f) # `Ints` = 150, `alpha` = 0.01

6 Let # `Ints` = 100. For the following sample sizes, calculate 95% confidence intervals for the population parameter. Print each graph. As the sample sizes increase, what happens to the widths of the confidence intervals? Is the percentage of times the confidence intervals capture the true value of the population proportion affected by these changing sample sizes? For which cases are the assumptions violated? What do you notice about the observed number of intervals containing the true value of the parameter for these cases? How large does the sample size(s) need to be to satisfy the assumptions?

   (a) n_1 = 5, (n_2 = 5)   (b) n_1 = 7, (n_2 = 7)   (c) n_1 = 10, (n_2 = 10)
   (d) n_1 = 15, (n_2 = 15)   (e) n_1 = 25, (n_2 = 25)   (f) n_1 = 50, (n_2 = 50)

12

Calculating Confidence Intervals

In Chapter 11, we saw how confidence intervals are interpreted. In this chapter, we give some examples of calculating them for real data.

12.1 Introduction

This lab brings together all of the one- and two-sample confidence intervals for means, variances, and proportions typically covered in introductory statistics courses. See the *Interpreting Confidence Intervals* Lab for a review of confidence intervals.

12.2 Objectives

This lab is primarily computational. It makes it easy to perform tests and to interpret the results.

12.3 Description

Given summary statistics (sample means, variances, proportions) for a confidence interval problem, the lab is used as follows. To begin the lab, select `Calculating Confidence Intervals` from the `Labs` menu. The dialog box shown in Figure 12.1 will appear; this box contains a number of items:

FIGURE 12.1
Dialog box for the *Calculating Confidence Intervals* Lab.

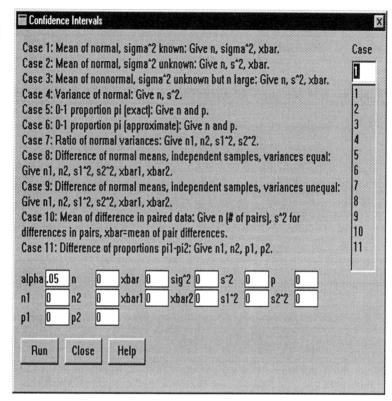

1. A list of 11 confidence interval *cases* usually taught in introductory statistics courses. At the end of this chapter are the formulas needed for the cases. For each case, the dialog box specifies what input is required.

2. A box labeled Case, where you select the number of the desired case. The initial entry is 1.

3. A set of boxes where you enter required input, including sample size(s), mean(s), variance(s), proportion(s), and 1 minus the level of confidence (the initial value is $\alpha = 0.05$).

4. The usual three buttons:

   (a) Run: After selecting Case and filling in alpha and any other information required, click on Run to execute the lab. The numerical results of the lab are displayed in the results window, including a description of the case, the input supplied, and the lower and upper limits of the confidence interval.

   (b) Close: Clicking on this button closes the lab and returns you to the Stat-Concepts menus.

   (c) Help: Clicking on this button opens a help window containing information about the lab.

When the lab begins, all of the information required for the first example in our guided tour has already been entered into the dialog box. Thus, clicking on Run shows the analysis for that example.

12.4 Guided Tour of the Lab

In this section, we consider seven examples of calculating confidence intervals.

12.4.1 Example 1

In a study of the keyboard height that typists prefer, a sample of 31 trained typists was selected and their preferred heights determined. The resulting average height was 80 cm. Assuming that the population of all typists' preferred heights is normally distributed with standard deviation 2 cm, find a 95% confidence interval for the mean of the population.

Analysis of Example 1

Under the normality assumption, the fact that the population standard deviation is known means that this is Case 1, so $n = 31$, $\bar{X} = 80$, $\sigma^2 = 4$, and $\alpha = 0.05$ must be entered in the dialog box. The resulting confidence interval is (79.296, 80.704).

Results Window for Example 1

```
95% CI for Mean of normal, sigma known:
n = 31, sigma^2 = 4, xbar = 80
Lower Limit = 79.29596
Upper Limit = 80.70404
```

12.4.2 Example 2

Consider an industrial process producing units that are classified as either defective or nondefective. In a random sample of 200 units, 6 are found to be defective. Find a 95% confidence interval for the proportion of all units that are defective. Use both the exact and approximate methods. Is the approximate method valid in this instance?

Analysis of Example 2

The exact and approximate confidence intervals for a population proportion are Cases 5 and 6 in the dialog box, so the lab must be run twice. In both cases, you must enter $n = 200$ and $p = 6/200 = 0.03$. The two confidence intervals are listed below. Note that the approximate method is probably valid, because the rule is that the smaller of $n\pi$ and $n(1 - \pi)$ must be at least 5, and in this case, the smaller of np and $n(1 - p)$ is 6. Although this is no guarantee of validity, most statisticians would regard the smaller of np and $n(1 - p)$ as a reasonable proxy for the smaller of $n\pi$ and $n(1 - \pi)$, which we can't know unless we know π—which is what we're trying to estimate!

Results Window for Example 2

```
95% CI for 0-1 proportion pi (exact):
n = 200, p = .03, X = 6
Lower Limit = .01108755
Upper Limit = .06415077

95% CI for 0-1 proportion pi (approximate):
n = 200, p = .03
Lower Limit = .00635825
Upper Limit = .05364175
```

12.4.3 Example 3

To compare two teaching methods, a school found 45 students with similar backgrounds and randomly selected 25 students to receive the first method and 20 students to recieve the second. At the end of the school year, the same test was administered to both groups and the results recorded as follows: $\bar{X}_1 = 75$, $\bar{X}_2 = 79$, $s_1^2 = 9$, and $s_2^2 = 16$.

1. Assuming that the populations of test scores are normally distributed, use a 95% confidence interval to decide if it is reasonable to assume that the variability in the two populations is the same.
2. Based on your result, use a 95% confidence interval to decide if it is reasonable to conclude that the two testing methods lead to the same average test score.

Analysis of Example 3

The first part of this problem involves comparing the variances of two normal samples. This is Case 7, so we enter $\alpha = 0.05$, $n_1 = 25$, $n_2 = 20$, $s_1^2 = 9$, and $s_2^2 = 16$, and find the 95% confidence interval for the ratio of the two variances to be (0.229, 1.319), which includes the value 1. This means it is reasonable to conclude that the ratio of the variances is 1, that is, that the two variances are the same.

Given that the variances are equal, the second part of the problem is Case 8. Therefore, we enter the same quantities as in the first part, as well as $\bar{X}_1 = 75$ and $\bar{X}_2 = 79$. The confidence interval for the difference between means is $(-6.104, -1.896)$, which does not include 0 as a reasonable value. Thus, we conclude that the two population means are different.

Results Window for Example 3

```
95% CI for Ratio of normal variances:
n1 = 25, n2 = 20, s1^2 = 9, s2^2 = 16
Lower Limit = .2293752
Upper Limit = 1.3191479

95% CI for Difference of normal means,
  independent samples, variances equal:
n1 = 25, n2 = 20, xbar1 = 75, xbar2 = 79,
  s1^2 = 9, s2^2 = 16
Lower Limit = -6.1039156
Upper Limit = -1.8960844
```

12.4.4 Example 4

A savings and loan association that finances many home purchases in a particular region wanted information on the extent to which mortgage payments reduced the amount of disposable income during the initial year of occupancy for first-time home buyers. A sample of 35 recently granted mortgage applications was obtained, and for each, the amount of the monthly mortgage payment as a percentage of take-home income was computed. The sample average percentage was found to be 24.7, with a standard deviation of 3.2. Assuming that percentage is a normally distributed variable (among all applicants), compute a 90% confidence interval for the true average percentage.

Analysis of Example 4

Given the normality assumption and the fact that we don't know the population variance, this is Case 2, so we enter $\alpha = 0.10$, $n = 35$, $\bar{X} = 24.7$, and $s^2 = 10.24$. Note that the problem gives $s = 3.2$, so we have to square s to get $s^2 = 10.24$ before we can fill in the dialog box. We can use the StataQuest calculator to do this. The resulting 90% confidence interval is (23.785,25.615).

Results Window for Example 4

```
90% CI for Mean of normal, sigma unknown:
n = 35, s^2 = 10.24, xbar = 24.7
Lower Limit = 23.785381
Upper Limit = 25.614619
```

12.4.5 Example 5

A sample of 250 petitioners who filed for bankruptcy between 1964 and 1967 waited an average of 35.41 months to clear the debts listed on the petitions, with a sample standard deviation of 21.34. Find a 99% confidence interval for the average number of months for all such petitioners.

Analysis of Example 5

This is similar to the previous example except that we don't know whether the population is normal. It is still Case 2, however, because the sample size is large. Thus, we enter $\alpha = 0.01$ and $\bar{X} = 35.41$, and use the StataQuest calculator to find that $s^2 = 21.24^2 = 455.3956$. The resulting 99% confidence interval is (31.907,38.913). Note that this is a fairly wide interval because there is so much variability in the wait times of the petitioners.

Results Window for Example 5

```
99% CI for Mean of normal, sigma unknown:
n = 250, s^2 = 455.3956, xbar = 35.41
Lower Limit = 31.906663
Upper Limit = 38.913337
```

12.4.6 Example 6

A test applied to samples of sand led to the following results:

```
26.7 25.8 24.0 24.9 26.4 25.9 24.4 21.7
24.1 25.9 27.3 26.9 27.3 24.8 23.6
```

Assuming the normality assumption is reasonable, find a 95% confidence interval for the mean result of the test for all sand.

Analysis of Example 6

Once again, this is Case 2 because of the normality assumption and unknown σ^2. In this example, we are given only the original data. Therefore, before filling in the dialog box, we must (1) use the StataQuest editor to enter the data and (2) use Summaries⟶ Means and SDs to find $\bar{X} = 25.1333$ and s. We then use the calculator to find $s^2 = 2.4927$. The 95% confidence interval is (24.259,26.008).

Results Window for Example 6

```
95% CI for Mean of normal, sigma unknown:
n = 15, s^2 = 2.4927, xbar = 25.1333
Lower Limit = 24.258974
Upper Limit = 26.007626
```

12.4.7 Example 7

The following are data from a study on the amount of stress experienced by a certain class of workers:

```
14.70 15.10 16.90 17.40 19.00 20.00 20.30
21.00 21.90 22.00 22.90 23.80 23.90 24.80
25.00 25.80 27.00
```

Assuming the data are normally distributed, find a 95% confidence interval for the standard deviation of the population of stresses.

Analysis of Example 7

This example is similar to the previous one except that it asks for a confidence interval for σ. Case 4 involves confidence intervals for σ^2 for normal populations. Thus, enter $\alpha = 0.05$ and $n = 17$, and obtain $s^2 = 13.7324$ by again entering the data, calculating the sample mean and standard deviation, and then squaring s. The lab gives the 95% confidence interval for σ^2 of (7.494,32.894). Using the calculator again to find the square root of these two limits gives the 95% confidence interval for σ of (2.737,5.735).

Results Window for Example 7

```
95% CI for Variance of normal:
n = 17, s^2 = 13.7324
```

```
Lower Limit = 7.4937264
Upper Limit = 32.893515

. di sqrt(7.4937)
2.7374623

. di sqrt(32.8935)
5.7352855
```

12.5 Summary

This lab makes it easy to find confidence intervals for the situations usually covered in introductory statistics courses.

12.6 Confidence Interval Formulas

Case 1: Normal mean μ (σ^2 known) (`1 sample normal`)

$$\bar{X} \pm Z_{\alpha/2} \frac{\sigma}{\sqrt{n}}$$

Case 2: Normal mean μ (σ^2 unknown) (`1 sample t`)

$$\bar{X} \pm t_{\alpha/2, \nu} \frac{s}{\sqrt{n}}, \quad \nu = n - 1$$

Case 3: Non-normal mean μ (σ^2 unknown, n large) (`1 sample t`): The formulas are the same as for Case 2.

Case 4: Normal variance σ^2 (`1 sample variance`)

$$\left(\frac{(n-1)s^2}{\chi^2_{\alpha/2, \nu}}, \frac{(n-1)s^2}{\chi^2_{1-\alpha/2, \nu}} \right), \quad \nu = n - 1$$

Case 5: Proportion π (exact CI for small n)

$$\left(\frac{X F_{1-\alpha/2, \nu_1, \nu_2}}{(n - X + 1) + X F_{1-\alpha/2, \nu_1, \nu_2}}, \frac{(X+1) F_{\alpha/2, \nu_3, \nu_4}}{(n - X) + (X + 1) F_{\alpha/2, \nu_3, \nu_4}} \right),$$

$X = np, \quad \nu_1 = 2X, \quad \nu_2 = 2(n - X + 1), \quad \nu_3 = 2(X + 1), \quad \nu_4 = 2(n - X)$

Case 6: Proportion π ($\min(n\pi, n(1-\pi)) \geq 5$) (`1 sample proportion`)

$$p \pm Z_{\alpha/2} \sqrt{p(1-p)/n}$$

Case 7: Ratio of normal variances (independent samples) (2 sample variances)

$$\left(\frac{s_1^2/s_2^2}{F_{\alpha/2,\nu_1,\nu_2}}, \frac{s_1^2/s_2^2}{F_{1-\alpha/2,\nu_1,\nu_2}} \right), \quad \nu_1 = n_1 - 1, \quad \nu_2 = n_2 - 1$$

Case 8: Difference of two normal means (σ_1^2, σ_2^2 unknown, but equal, independent samples) (2 sample t)

$$(\bar{X}_1 - \bar{X}_2) \pm t_{\alpha/2,\nu} \sqrt{s_P^2(1/n_1 + 1/n_2)}, \quad s_P^2 = \frac{(n_1 - 1)s_1^2 + (n_2 - 1)s_2^2}{n_1 + n_2 - 2},$$
$$\nu = n_1 + n_2 - 2$$

Case 9: Difference of two normal means (σ_1^2, σ_2^2 unknown, unequal, independent samples) (2 sample t)

$$(\bar{X}_1 - \bar{X}_2) \pm t_{\alpha/2,\nu} \sqrt{s_1^2/n_1 + s_2^2/n_2}, \quad \nu = \frac{\left(s_1^2/n_1 + s_2^2/n_2\right)^2}{\dfrac{\left(s_1^2/n_1\right)^2}{n_1 - 1} + \dfrac{\left(s_2^2/n_2\right)^2}{n_2 - 1}}$$

Case 10: Mean of differences in normal paired data (σ^2 unknown) (paired t)

$$\bar{d} \pm t_{\alpha/2,\nu} \frac{s_d}{\sqrt{n}}, \quad \nu = n - 1, \quad \bar{d} \text{ and } s_d \text{ are mean and standard deviation of differences in pairs.}$$

Case 11: Difference $\pi_1 - \pi_2$ of proportions (independent samples, $\min(n_1\pi_1, n_1(1 - \pi_1)) \geq 5$, $\min(n_2\pi_2, n_2(1 - \pi_2)) \geq 5$) (2 sample proportions)

$$(p_1 - p_2) \pm Z_{\alpha/2} \sqrt{\frac{p_1(1 - p_1)}{n_1} + \frac{p_2(1 - p_2)}{n_2}}$$

13

Tests of Significance

In the *Calculating Confidence Intervals* and *Interpreting Confidence Intervals* Labs, we saw how confidence intervals can be used to study whether a certain value for a parameter is reasonable. For example, if someone suggests that the average height of six-year-olds is 48 inches, we can randomly select a sample of n six-year-olds, calculate $\bar{X}$ and s^2 for the sample, and then see if 48 inches is between $\bar{X} - t_{0.025, n-1} s/\sqrt{n}$ and $\bar{X} + t_{0.025, n-1} s/\sqrt{n}$. If it is, we can conclude that 48 inches is a reasonable value for μ; if it isn't, we can conclude (with a 5% chance of error) that 48 inches is not reasonable. We have studied such procedures for $\mu, \sigma^2, \mu_1 - \mu_2, \sigma_1^2/\sigma_2^2, \pi$, and $\pi_1 - \pi_2$.

13.1 Introduction

In this lab, we consider another way to test claims, namely, what are called *tests of significance*. This method is again based on the fact (discussed extensively in the *Z, t, χ^2, and F* and *Sampling Distributions* Labs) that the Z statistic

$$Z = \frac{\bar{X} - \mu}{\sigma/\sqrt{n}}$$

and the one-sample t statistic

$$t = \frac{\bar{X} - \mu}{s/\sqrt{n}}$$

should have values close to 0 if $\bar{X}$ is close to μ. Also, the two-sample t statistic

$$t = \frac{(\bar{X}_1 - \bar{X}_2) - (\mu_1 - \mu_2)}{\sqrt{s_p^2(1/n_1 + 1/n_2)}},$$

where

$$s_p^2 = \frac{(n_1 - 1)s_1^2 + (n_2 - 1)s_2^2}{n_1 + n_2 - 2},$$

should have values close to 0 if the difference between the sample means is close to the difference between the population means. Further, the chi-square statistic

$$\chi^2 = \frac{(n-1)s^2}{\sigma^2}$$

should be close to $(n-1)$ if the sample variance s^2 is close to the population variance σ^2. Finally, the F statistic

$$F = \frac{s_1^2/\sigma_1^2}{s_2^2/\sigma_2^2}$$

should be close to 1 if the sample variances of two samples are close to the population variances of the two populations they were sampled from.

Further, the sampling distribution of the Z statistic is the Z curve; the sampling distributions of the one- and two-sample t statistics are the t curves with $n - 1$ and $n_1 + n_2 - 2$ degrees of freedom, respectively; and the sampling distributions of the χ^2 and F statistics are the χ^2 curve with $n - 1$ degrees of freedom and the F curve with numerator and nominator degrees of freedom $n_1 - 1$ and $n_2 - 1$.

Thus, to test a claimed value of a parameter, we substitute the hypothesized value into the appropriate statistic and then see how far the statistic is from what it should be if the claim is true. More specifically, we find the probability that the statistic is as far away from what it should be (or farther) if the claim is true. This probability is called the *p value* of the test of significance.

In this lab, we study how to find such probabilities if we have already calculated the value of the statistic. You will have a choice of telling the lab the value or having the lab give you a value.

13.1.1 Types of Tests

There are three kinds of claims we can make about a parameter:

1 *Right-sided alternative:* In this case, the claim is that the parameter is *at most* some value. Thus, the opposite of the claim is that the parameter is *greater than* the claimed value, and we would calculate the probability that the statistic is greater than the observed value. This would be an area in the *right tail* of either the Z, t, χ^2, or F curve, depending on which situation we're studying, and so this test is called a right-sided test (see Figure 13.1). In the figure, the value of the Z statistic turns out to be 0.794. Thus, the p value is the probability that Z is greater than 0.794, which is 0.214.

FIGURE 13.1
Example of a *p* value for a right-sided *Z* test.

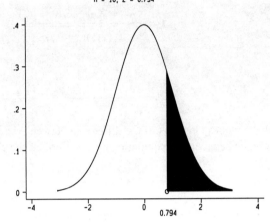

2. *Left-sided alternative:* This is the opposite of the right-sided test, so the claim is that the parameter has *at least* some value, and we end up finding a probability in the *left tail* of a distribution curve (see Figure 13.2).

FIGURE 13.2
Example of a *p* value for a left-sided *t* test.

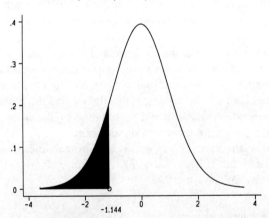

3. *Two-tailed alternative:* In this case, the claim is that the value of the parameter is *equal to* some value, with the alternative being that the value is either *less than* or *greater than* the claimed value. Thus, we find the *p* value as the sum of areas in *both tails* of the appropriate curve (see Figure 13.3). Note that the statistic determines one of the shaded areas (on the left if the statistic is to the left of the expected value and on the right if the statistic is to the right of the expected value), while the other shaded region is chosen to have the same area as the first region. In Figure 13.3, the χ^2 statistic is 3.835 (you can tell because that is where the little circle marking the observed value is), which is to the left of the expected

FIGURE 13.3
Example of a p value for a two-sided χ^2 test.

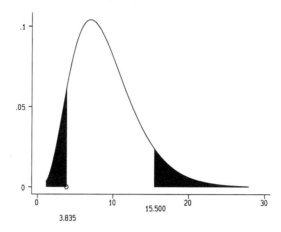

value of 9 (which is $n - 1$). Thus, the p value is 0.156, which is twice the area under the χ_9^2 curve to the left of 3.835. The lab also shades a region in the right tail having the same area as the region on the left.

We will look at these graphs again in our guided tour.

13.2 Objectives

1. Show how to find p values for right-sided, left-sided, and two-sided tests of significance for the Z, one- and two-sample t, χ^2, and F situations.
2. Illustrate graphically what p values represent.

13.3 Description

To start the lab, select Tests of Significance from the Labs menu. The dialog box shown in Figure 13.4 will appear containing information about the lab, as well as six boxes and three buttons. You choose which situation you want to study (the Z, one-sample or two-sample t, χ^2, or F), what test you want (right-sided, left-sided, or two-sided), and whatever sample size (or sizes if you've chosen the two-sample t or the F situations) you want. You also can choose whether you want to supply a value for the transformed statistic. If you want the lab to generate a value, then it randomly generates a sample (or samples) from a normal population having mean 0 and variance 1, calculates the value of the transformed statistic, and draws the appropriate curve. It then shades in the tail (or tails for a two-sided test) of the curve and displays the value of the transformed statistic and the p value above the plot.

The dialog box contains these six boxes:

1. n_1: If you want to do a Z, one-sample t, or χ^2 test, then you supply the sample size in this box. If you want to do a two-sample t or F test, this is where you

FIGURE 13.4
Dialog box for the *Tests of Significance* Lab.

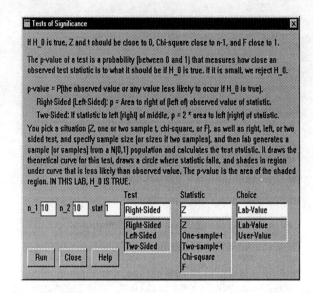

enter the size of the first sample. The initial value is 10, and the allowed values are from 2 to 100.

2 `n_2`: For the two-sample *t* and *F* tests, this is where you enter the second sample size. For the other tests, simply ignore this box. Again, the initial value is 10, and allowed values are from 2 to 100.

3 `stat`: If you want to supply your own value of the test statistic, this is where you enter it.

4 `Test`: This is where you specify whether you want a right-sided, left-sided, or two-sided test.

5 `Statistic`: This is where you specify which test statistic you want.

6 `Choice`: This is where you choose whether you want the lab to supply a value of the test statistic or you want to supply your own.

In the bottom left-hand corner of the dialog box are three buttons:

1 `Run`: Clicking on this button executes the lab.

2 `Close`: Clicking on this button closes the lab and returns you to the StatConcepts menus.

3 `Help`: Clicking on this button opens a help window containing information about the lab.

Note that in this lab, the "claimed values" of the parameters are as follows:

1 Z: $\mu = 0$
2 One-sample *t*: $\mu = 0$
3 Two-sample *t*: $\mu_1 = \mu_2 = 0$
4 χ^2: $\sigma^2 = 1$
5 F: $\sigma_1^2 = \sigma_2^2 = 1$

Thus, in all cases in this lab, the claimed value is the true value. This means that if the lab is supplying the value of the test statistic, then most of the time its value will be close to what it should be (0 for Z and t, $(n-1)$ for χ^2, and 1 for F). Therefore, the p values should not be extremely small.

13.4 Guided Tour of the Lab

We begin our tour by looking more closely at Figures 13.1–13.3. In Figure 13.1 we chose to have the lab supply a value of the Z statistic for a right-sided test for $n = 10$. The value of Z was 0.794, which is to the right of the value 0 and which is what Z should be if the claim is true. The question is whether it is too far to the right of 0 to have the claim be reasonable. The statistical answer to this question involves finding the area to the right of 0.794 under the Z curve. We look to the right because we're doing a right-sided test and we look at the Z curve because we're using the Z statistic. The shaded area is 0.214, which is not small (usually the p value must be less than a number such as 0.05 or 0.025 or 0.01 to call into question the reasonableness of a claim), and so we cannot conclude that the claim that the population mean is 0 is unreasonable.

In Figure 13.2, we selected a left-sided two-sample t test with sample sizes $n_1 = n_2 = 10$, and the lab selected the value for the t statistic of -1.144. Thus, the p value is the area to the *left* of -1.144 under the t curve having $n_1 + n_2 - 2 = 18$ degrees of freedom. This area is shown in the figure as 0.134, which, again, is not small enough to doubt the reasonableness of the claim that the means of the two populations are the same.

In Figure 13.3, the lab supplied the value 3.835 for a χ^2 transformed statistic, and we chose a two-sided test with $n = 10$. Finding the p value for this case is more complicated than in the previous two cases because (1) the χ^2 curve is not symmetrical while the Z and t curves are, and (2) we need to find areas in both tails of the curve and add them together. But this task can be simplified by following the *basic rule of p values for two-sided tests*.

13.4.1 Finding p Values for Two-sided Tests

No matter which situation (Z, t, χ^2, or F) you are working with:

1. If the value of the transformed statistic is *less than* it should be if the claim is true (0 for Z and t, $n-1$ for χ^2, and 1 for F), then the p value is twice the area to the *left* of the value.

2. If the value of the transformed statistic is *greater than* it should be if the claim is true (0 for Z and t, $n-1$ for χ^2, and 1 for F), then the p value is twice the area to the *right* of the value.

In Figure 13.3, the transformed statistic is 3.835, which is less than $n - 1 = 9$, and so we find the area to the left of 3.835, which is 0.078. To get the p value, we double this and get 0.156. In the figure, we have shaded in the area under the left side of the χ^2 curve to the left of 3.835 and also a region on the right that has the same probability content as the one on the left, namely, 0.078.

13.4.2 p Values As Statistics Get More Extreme

We conclude our brief tour by looking at the F situation as the value of the F transformed statistic gets farther away from the expected value of 1 if the claim of equal population variances is true. We consider a two-sided test with $n_1 = n_2 = 20$, and we run the lab twice, first for $F = 1.75$ and then for $F = 2.25$. The resulting graphs are shown in Figures 13.5 and 13.6. Notice what happens. The value 1.75 is to the right of 1, so the p value is twice the area under the F curve to the right of 1.75, and thus is 0.232. When we increase F to 2.25, it is even farther to the right of 1, so the shaded area gets smaller, namely, 0.0425 on each side, giving the p value of 0.085. If we continued to let F get more extreme (farther away from 1), the shaded area would continue to get smaller.

See the chapter exercises for some more ways you can use this lab.

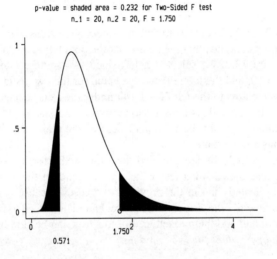

FIGURE 13.5 Example of a p value for a two-sided F test.

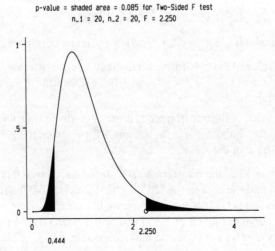

FIGURE 13.6 Example of a p value for a two-sided F test.

13.5 Summary

We have seen the following in this lab:

1. We can find p values for right-sided, left-sided, and two-sided tests of significance for the Z, one- and two-sample t, χ^2, and F situations.
2. For a two-sided test, we simply double the area in the tail of the curve on the side of the expected value for the test where the statistic falls.
3. The smaller the p value is, the more we call into question the reasonableness of the claim being made.

13.6 Lab Exercises

1. Run the lab for each of the following right-sided situations. Print the graphs and make a table of p values. What are the claims and alternatives in each situation? In which situations would you doubt the reasonableness of the claim being made?
   (a) Z with $Z = 1.5$
   (b) One-sample t with $n = 10$ and $t = 2.0$
   (c) Two-sample t with $n_1 = 10$, $n_2 = 20$, and $t = 2.5$
   (d) Chi-square with $n = 10$ and $\chi^2 = 15$
   (e) F with $n_1 = 30$, $n_2 = 50$, and $F = 1.5$

2. Run the lab for each of the following left-sided situations. Print the graphs and make a table of p values. What are the claims and alternatives in each situation? In which situations would you doubt the reasonableness of the claim being made?
   (a) Z with $Z = -1.5$
   (b) One-sample t with $n = 10$ and $t = -2.0$
   (c) Two-sample t with $n_1 = 10$, $n_2 = 20$, and $t = -2.5$
   (d) Chi-square with $n = 10$ and $\chi^2 = 3$
   (e) F with $n_1 = 30$, $n_2 = 50$, and $F = 0.4$

3. Run the lab for each of the following two-sided situations. Print the graphs and make a table of p values. What are the claims and alternatives in each situation? In which situations would you doubt the reasonableness of the claim being made? In which situations are the two shaded areas *symmetrical*?
   (a) Z with $Z = 1.5$
   (b) One-sample t with $n = 10$ and $t = 2.0$
   (c) Two-sample t with $n_1 = 10$, $n_2 = 20$, and $t = 2.5$
   (d) Chi-square with $n = 10$ and $\chi^2 = 15$
   (e) F with $n_1 = 30$, $n_2 = 50$, and $F = 1.5$

4. Run the lab 20 times for a right-sided Z test with $n = 10$. Each time record the p value you obtain. Do the values seem to be well spread between 0 and 1? (This

is what should happen.) How many of the 20 p values are less than 0.05? How many should be?

5 Do the same thing as in the previous exercise except for the two-sided F test with $n_1 = n_2 = 20$. Record the values of the statistic and the p value. Do the values of F seem to fall around $F = 1$?

6 Under the following conditions, perform an experiment similar to that in the last part of the guided tour. Print the resulting graphs and make a table of the p values. What is happening to the shaded areas as the statistic gets farther from the expected value if the claim being made is true?

   (a) Two-sided Z test with $Z = 1.0, 1.5, 2.0$, and 2.5
   (b) Right-sided one-sample t test with $n_1 = 10$ and $t = 1.0, 1.5, 2.0$, and 2.5
   (c) Right-sided two-sample t test with $n_1 = n_2 = 10$ and $t = 1.0, 1.5, 2.0$, and 2.5
   (d) Two-sided chi-square test with $n = 20$ and $\chi^2 = 25, 28, 31$, and 34
   (e) One-sided F test with $n_1 = 40$ and $n_2 = 10$ and $F = 0.8, 0.7, 0.6$, and 0.5

7 What happens to the p value for the two-sided F test with $F = 1.5$ if we increase $n_1 = n_2$ from 20 to 50? Print the two graphs and write down the two p values. Why does this happen?

8 Use StataQuest to verify that the p values in Figures 13.1–13.3 are correct. (Use the `Statistical Tables` item on the `Calculator` menu.)

14

Level of Significance of a Test

Many times, a change is introduced into a population, and the effect of that change is of interest. For instance, it might be of interest to see how reducing the fat in a person's diet will reduce his or her mean blood pressure, or how a scandal changes the proportion of voters who will vote for a particular candidate, or how increasing the speed of a production line will decrease the quality of a product. In any case, after data have been appropriately collected, a hypothesis test could be performed to answer these questions.

14.1 Introduction

In the *Sampling Distributions* Lab, we studied five statistics and their sampling distributions under specific assumptions. We also saw how the sampling distributions were affected when the assumptions were not met. In this lab, we will study how we can use these sampling distributions to draw conclusions in hypothesis testing and how probabilistic statements can be made about the certainty of those decisions.

14.1.1 Some Basic Ideas

All hypothesis tests have the same basic structure:

1. A null hypothesis H_0 and an alternative hypothesis H_a.
2. A test statistic. In the *Sampling Distributions* Lab they were Z, one-sample t, two-sample t, χ^2, or F, and were classified in the following manner:

(a) One-sample inference procedures on the population mean μ:

   i *Case 1*: The Z statistic.

   ii *Case 2*: The (one-sample) t statistic.

(b) *Case 3*: The two-sample inference procedure on the difference between two population means $\mu_1 - \mu_2$; two-sample t statistic.

(c) *Case 4*: The one-sample inference procedure on the population variance σ^2; the χ^2 statistic.

(d) *Case 5*: The two-sample inference procedure on the ratio of two population variances σ_1^2/σ_2^2; the F statistic.

3 A rejection rule based on the sampling distribution of the test statistic (either in the form of a rejection region or a p value).

4 A decision in terms of the null hypothesis ("reject H_0" or "fail to reject H_0").

When performing a hypothesis test, we have four possible realities:

1 The null hypothesis H_0 is true and we fail to reject H_0, in which case a correct decision was made.

2 The null hypothesis H_0 is true and we reject H_0. This is known as a *Type I error*. The probability of a Type I error is denoted by α and is called the *level of significance* of the test.

3 The null hypothesis H_0 is false and we fail to reject H_0. This is known as a *Type II error*. The probability of a Type II error is denoted by β. The *power of the test* is equal to $1 - \beta$. We will study the power of the test in the *Power of a Test* Lab (Chapter 16).

4 The null hypothesis H_0 is false and we reject H_0, in which case a correct decision was made.

In this lab, we are studying the level of significance of a test, so, the null hypothesis will always be true.

Of course, in practice, we don't know whether the null hypothesis is true. But if we know the sampling distribution of the test statistic, we can use that distribution to determine how likely our decision is to result in an error. The conclusion of the hypothesis test determines what type of error we might be making. If we reject H_0, we could be making a Type I error; if we fail to reject H_0, we could be making a Type II error. In this lab, we will study the probability of making a Type I error.

The value of α should be interpreted very carefully. Suppose the null hypothesis is true. Then, if we could take millions and millions of samples of size n from the population, α is the proportion of samples that would result in rejecting that (true) null hypothesis.

14.2 Objectives

1 Learn the definition of a Type I error.
2 Properly interpret the level of significance of a test.

3 For each of the five test statistics presented in the lab, learn which violation of assumptions affects the reliability of the statistical inference in terms of the probability of a Type I error.

14.3 Description

To begin the lab, choose `Level of Significance of a Test` from the `Labs` menu. A dialog box like the one shown in Figure 14.1 opens containing a brief description of the lab. Below that are six small boxes requiring information.

1. `n_1`: The number supplied in the box to the immediate right of `n_1` indicates the size of the samples to be chosen from the population if `Statistic` is any of `Z`, `One-sample-t`, or `Chi-square`. If `Statistic` is `Two-sample-t` or `F`, then `n_1` is the size of the samples drawn from the first population. The value of `n_1` should be an integer and must be between 2 and 100. The initial value of `n_1` is 10.

2. `n_2`: The number supplied in the box to the immediate right of `n_2` indicates the size of the samples to be chosen from the second population if `Statistic` is `Two-sample-t` or `F`. If `Statistic` is `Z`, `One-sample-t`, or `Chi-square`, the value of `n_2` is ignored. The value of `n_2` should be an integer and must be between 2 and 100. The initial value of `n_2` is 10.

3. `alpha`: The number supplied in the box beneath `alpha` is the desired level of significance of the test. The value of `alpha` must be between, but not equal to, 0 and 1. If a value of `alpha` is desired that is not given in the choices, simply type in the desired value.

4. `Test`: The choice specified in the box beneath `Test` indicates the direction of the inequality in the alternative hypothesis:

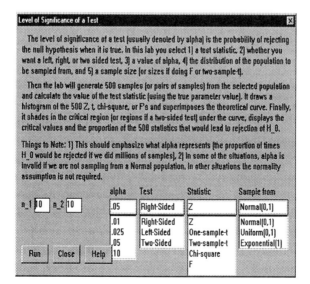

FIGURE 14.1
Dialog box for the *Level of Significance of a Test* Lab.

(a) If Test is Right-Sided, the alternative hypothesis is of the form $H_a: \theta > \theta_0$.

(b) If Test is Left-Sided, the alternative hypothesis is of the form $H_a: \theta < \theta_0$.

(c) If Test is Two-Sided, the alternative hypothesis is of the form $H_a: \theta \neq \theta_0$.

The initial Test is Right-Sided. The choices of Test are restricted to those shown.

5 Statistic: The choice specified for Statistic corresponds to the type of test that is to be performed as outlined in the five cases given in Subsection 14.1.1. The initial Statistic chosen for study is the Z statistic. The choices of Statistic are restricted to those given.

6 Sample from: The choice specified in the box beneath Sample from indicates the distribution of the population from which the samples are chosen. There are three distributions that may be selected: Normal(0,1), Uniform(0,1), and Exponential(1). All of these distributions are discussed in the *How Are Populations Distributed?* Lab. The initial distribution is Normal(0,1). The choices for Sample from are limited to those shown.

In the bottom left-hand corner of the dialog box are three buttons.

1 Run: Clicking on this button begins the lab. A graphics window will open that briefly displays a status bar that fills with yellow as 500 samples of size n are drawn from the population specified in Sample from. Then the value of the statistic specified in Statistic is calculated for each sample. After the samples are selected and computations are complete, a graph similar to the one shown in Figure 14.2 will open. This graph contains a yellow histogram of the 500 values of the test statistic with the theoretical sampling distribution superimposed in red. The region shaded in red has area equal to alpha and is the rejection region of the hypothesis test. At the top of the graph is a heading that contains information specifying the conditions under which the lab was run and the proportion of times that the null hypothesis was rejected for the 500 test statistics.

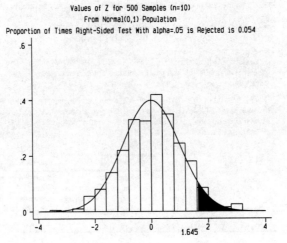

FIGURE 14.2
Rejection region for $\alpha = 0.05$ for a right-sided Z test based on 500 samples of size $n = 10$ from a Normal(0,1) population.

2 Close: Clicking on this button closes the lab and returns you to the StatConcepts menus.

3 Help: Clicking on this button opens a help window containing information about the lab.

14.4 Guided Tour of the Lab

The tour we are about to embark on contains many elaborate, intertwined facets, all of which have one thing in common: for all variations of the lab, the null hypothesis will be true. This will allow us to examine relationships between features unique to Type I errors—rejecting a true null hypothesis.

14.4.1 The Z Test Statistic

The Z statistic is used for inference procedures on the population mean. In this lab, because the null hypothesis $H_0: \mu = \mu_0$ is true, then the value of the mean μ is equal to the hypothesized value μ_0. Therefore, in the Z statistic, we can replace μ by its value μ_0, yielding the Z test statistic

$$Z = \frac{\bar{X} - \mu_0}{\sigma/\sqrt{n}}.$$

In the *Sampling Distributions* Lab, we learned that the Z statistic has a standard normal distribution if the parent population belongs to the family of normal distributions. If the parent population is not normally distributed, then the Central Limit Theorem ensures the sampling distribution of Z can still be well approximated by a standard normal distribution if the sample size n is large enough. Throughout our tour of the *Level of Significance of a Test* Lab, we are going to learn how the validity of these statements affects the actual proportion of samples that will result in a value of Z that rejects the null hypothesis and compare that proportion to the desired value of α.

Begin the lab by choosing Level of Significance of a Test from the Labs menu. Run the lab for n_1 = 10, alpha = 0.05, Test of Right-Sided, Statistic of Z, and Sample from of Normal(0,1). When you click on Run, you will see a yellow status bar. As the bar is filling, the lab is generating 500 samples from the Normal(0,1) population distribution. For each of the samples, the value of Z is calculated.

When the calculations are complete, the graph you see should look similar to the one shown in Figure 14.2. The graphs will differ somewhat in appearance because the lab generates a different set of 500 samples every time it is run. In this graph, the theoretical standard normal distribution is drawn in red. The shaded tail has area alpha = 0.05. The value 1.645 is the critical value $Z_{0.05}$ that cuts off the upper tail so that the area to the right is alpha = 0.05. (For more on this, see the *Critical Values* Lab.) The yellow histogram is of the 500 values of Z. The heading contains information about the values that were specified to run the lab. The last piece of information in the heading is the actual proportion of times that the null hypothesis was rejected for the 500 samples. In our figure, the proportion of samples that resulted in a value of Z larger than 1.645 (so that the null hypothesis was rejected) is 0.054. You should not be surprised by the closeness of the histogram to the theoretical curve, nor by the closeness of 0.054 to alpha = 0.05. For this run of the lab, the assumptions were satisfied, and the samples behaved exactly as expected.

Change Test to Left-Sided and run the lab again. The resulting graph should look similar to the one shown in Figure 14.3. You might recall from the *Critical Values* Lab that the critical value for a left-sided Z test is the negative of the critical value for the right-sided Z; that is, $Z_{0.95} = -1.645$. This is the only major feature of the graph that differs from the graph in Figure 14.2. In Figure 14.3, the proportion of the 500 samples that resulted in a value of Z less than -1.645 (so that the true null hypothesis was rejected) is 0.054. Again, because the assumptions were satisfied, the behavior of the samples is remarkably, but not surprisingly, close to the theoretical standard normal curve.

Now change Test to Two-Sided and run the lab again. The resulting graph should look similar to the one shown in Figure 14.4. The two tail areas shaded in red are equal and must sum to alpha = 0.05; therefore, they are each $0.05/2 = 0.025$. The two critical values are $Z_{0.025} = 1.96$ and $Z_{0.975} = -1.96$. The assumption that the population distribution is a member of the normal family is satisfied, so the agreement between the curve and the histogram is expected. For this run of the lab, the proportion of the 500 samples that resulted in a test statistic that was either large enough or small enough to reject H_0 was exactly 0.050 = alpha!

For a Right-Sided Test, change the value of alpha to 0.10, and run the lab. In Figure 14.5, we see the close agreement of the Normal(0,1) curve and the histogram of the 500 values of Z. The main difference between this graph and the one in Figure 14.2 is that the shaded portion of the curve now has area 0.10. The proportion of the 500 Z's that were large enough that the null hypothesis was rejected was 0.096, which is reasonably close to 0.10. Changing the level of significance will not affect the reliability of the result of the hypothesis test if the assumptions are satisfied. If you wish, you may also run the lab for alpha = 0.10 for a Left-Sided and

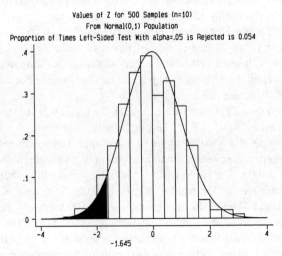

FIGURE **14.3**
Rejection region for $\alpha = 0.05$ for a left-sided Z test based on 500 samples of size $n = 10$ from a Normal(0,1) population.

FIGURE **14.4**
Rejection region for $\alpha = 0.05$ for a two-sided Z test based on 500 samples of size $n = 10$ from a Normal(0,1) population.

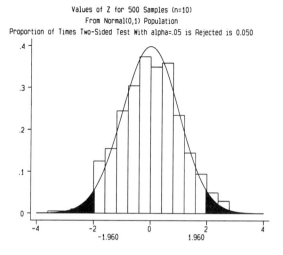

FIGURE **14.5**
Rejection region for $\alpha = 0.10$ for a right-sided Z test based on 500 samples of size $n = 10$ from a Normal(0,1) population.

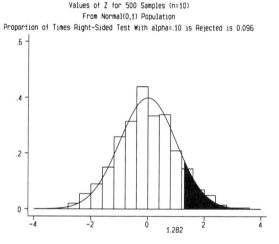

Two-Sided test. You will see the same goodness of results as you did for alpha = 0.05. Only the critical values and the tail areas will change.

Recall from the *Sampling Distributions* Lab that if the parent population is normally distributed, the sampling distribution of Z is unaffected by the change in the sample size. Regardless of n, the sampling distribution of Z is the standard normal distribution. Rather than examine that same point again, we will now focus on how changing the distribution of the parent population affects the proportion of samples that result in a value of Z that leads us to reject the true null hypothesis.

Let n_1 = 5, Test be Right-Sided, alpha = 0.05, and Sample from be Uniform(0,1). Run the lab. Another fact you may recall from the *Sampling Distributions* Lab is that if the parent population is uniformly distributed, a "large enough" sample size to justify using the standard normal distribution for the sampling distribution of Z is somewhere between $n = 5$ and 10. Thus, we are not surprised by the closeness of the histogram of the 500 sample means and the theoretical

Normal(0,1) curve in Figure 14.6. For this run of the lab, the proportion of samples that resulted in rejecting H_0 is 0.05, identical to `alpha` = 0.05.

Recall that any member of the family of uniform distributions is symmetrical about its mean. The standard normal distribution is also symmetrical about its mean. These two facts together imply that the closeness of `alpha` and the proportion of samples that will result in rejecting H_0 for left-sided tests will be the same as for right-sided tests. To verify this, run the lab for the same values of `n_1` and `alpha`, and change `Test` to `Left-Sided`. The result is shown in Figure 14.7. The only major feature that has changed is that the rejection region is on the other side of zero. And notice that the proportion of samples that resulted in rejecting H_0 is 0.052, which is very close to `alpha`. Because the goodness of the results is the same for both sides of the test, it will also be so for two-sided tests.

How does increasing the sample size affect the closeness of these results? Change `Test` back to `Right-Sided` and let `n_1` = 20. Figure 14.8 shows the result, which is comparable to what we obtained in Figure 14.6 for `n_1` = 5. Because the parent

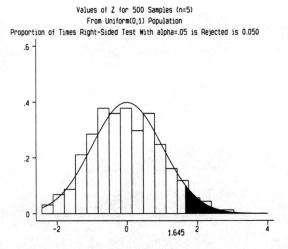

FIGURE 14.6
Rejection region for $\alpha = 0.05$ for a right-sided Z test based on 500 samples of size $n = 5$ from a Uniform(0,1) population.

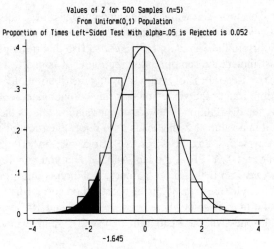

FIGURE 14.7
Rejection region for $\alpha = 0.05$ for a left-sided Z test based on 500 samples of size $n = 5$ from a Uniform(0,1) population.

FIGURE **14.8**
Rejection region for $\alpha = 0.05$ for a left-sided Z test based on 500 samples of size $n = 20$ from a Uniform(0,1) population.

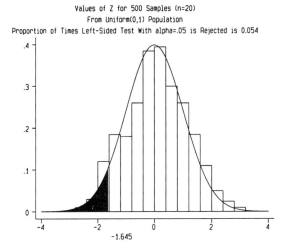

population has a uniform distribution, there is little need for even a moderately large sample size to move the sampling distribution of Z closer to the Normal(0,1) distribution. Small sample sizes are "large enough."

Change n_1 back to 5, and let Sample from be Exponential(1). We are now going to see how having a skewed parent distribution affects the closeness of the desired value of alpha and the actual proportion of Type I errors. Run the lab to obtain the graph shown in Figure 14.9. From the *Sampling Distributions* Lab, we know that the histogram of the 500 Z statistics will be skewed and so will not match the symmetrical Normal(0,1) curve. Examination of Figure 14.9 shows how this mismatch of distributions affects the proportion of samples, resulting in rejecting the null hypothesis. In our result, this proportion was 0.072, which is larger than the desired alpha = 0.05. This means that we are actually making a Type I error for approximately 7% of the samples instead of just 5%. The reliability of our inference is not as good as we would like.

FIGURE **14.9**
Rejection region for $\alpha = 0.05$ for a right-sided Z test based on 500 samples of size $n = 5$ from an Exponential(1) population.

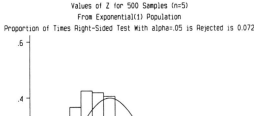
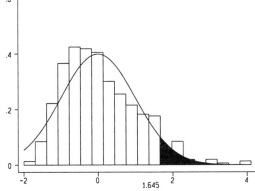

How is the left-sided Type I error rate affected? To see this, run the lab for `Test of Left-Sided` while leaving all other values the same. One possible result is shown in Figure 14.10. For the left-sided test, the proportion of samples that reject H_0 is much smaller than 0.05; it is 0.010. Here, the chance of making a Type I error is smaller than what is specified. This may seem encouraging, but in the *Power of a Test* Lab (Chapter 16), we will see that this is not the case. The uncertainty in the reliability of our results is evident with an exponentially distributed parent population.

How will increasing the sample size bolster our confidence in the reliability of our knowledge of whether we're making a Type I error? Change `Test` back to `Right-Sided` and let n_1 = 25. We can see in Figure 14.11 that increasing the sample size has helped greatly. The proportion of samples that resulted in rejecting H_0 for n_1 = 25 was 0.062, which is much closer to the desired 0.05. In the chapter exercises, you will also see how increasing the sample size affects the level of significance for left-sided and two-sided tests.

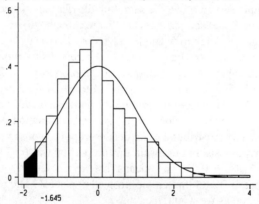

FIGURE 14.10
Rejection region for $\alpha = 0.05$ for a left-sided Z test based on 500 samples of size $n = 5$ from an Exponential(1) population.

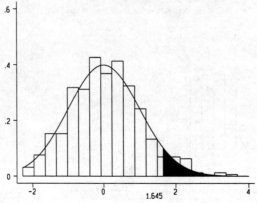

FIGURE 14.11
Rejection region for $\alpha = 0.05$ for a right-sided Z test based on 500 samples of size $n = 25$ from an Exponential(1) population.

14.4.2 The One-sample t Test Statistic

The one-sample t statistic is also used for purposes of statistical inference on the population mean μ. If we do not know the value of the population variance σ^2 and must estimate it with s^2, then the resulting test statistic

$$t = \frac{\bar{X} - \mu_0}{s/\sqrt{n}}$$

has a t_{n-1} distribution if the parent population is normally distributed. To examine the proportion of Type I errors for this statistic under a normal parent, let n_1 = 10, Test be Right-Sided, alpha = 0.05, and Sample from be Normal(0,1). Run the lab for these values.

Examine the resulting graph shown in Figure 14.12. The shaded rejection region still has area alpha = 0.05. Recall that the t_9 distribution has larger variance than the standard normal distribution, so that the tails of the t_9 distribution are higher or "thicker" than the Z distribution. This forces the critical value $t_{0.05,9} = 1.833$ to be larger than the Z for the same area. But note that this does not affect the closeness of alpha and the proportion of samples that rejected H_0, which was 0.044. If we ran the lab again for a left-sided test, we would see a similar result on the other side of zero, just as we did for the Z statistic. Two-sided tests would also exhibit the same agreement. Further, as long as the assumption of a normal parent distribution remains satisfied, the sampling distribution of the one-sample t statistic is exactly a t_{n-1} distribution, so the actual probability of a Type I error will be exactly the desired value of α; increasing the sample size n will only decrease the spread of the distribution.

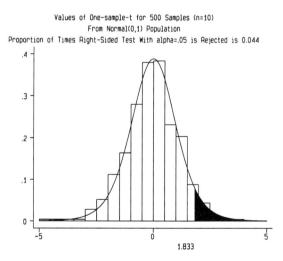

FIGURE 14.12 Rejection region for $\alpha = 0.05$ for a right-sided one-sample t test based on 500 samples of size $n = 10$ from a Normal(0,1) population.

The comparisons become more interesting when we deviate from a normally distributed parent. Let Sample from be Uniform(0,1) and Test be Right-Sided and run the lab. Because the parent distribution is a uniform distribution, we again see in Figure 14.13 that for this symmetrical t_9 distribution, the value of `alpha` and the proportion of samples rejecting H_0 are remarkably close. For a uniform parent distribution, a small sample size is large enough to confidently use the t distribution for purposes of statistical inference on μ.

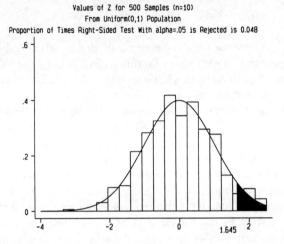

FIGURE **14.13**
Rejection region for $\alpha = 0.05$ for a right-sided one-sample t test based on 500 samples of size $n = 10$ from a Uniform(0,1) population.

Now change the distribution in `Sample from` to `Exponential(1)` and run the lab again. As Figure 14.14 shows, the true proportion of samples resulting in rejecting the null hypothesis obviously is much smaller than the desired 0.05. Change `Test` to `Left-Sided` to get a figure similar to the one shown in Figure 14.15. Now the chance of making a Type I error is quite a bit larger than the desired `alpha` = 0.05. In either case, the result is clear. The desired level of significance and the actual level of significance are not the same. Our statement of certainty in our inferential results is compromised.

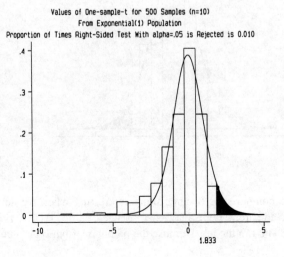

FIGURE **14.14**
Rejection region for $\alpha = 0.05$ for a right-sided one-sample t test based on 500 samples of size $n = 10$ from an Exponential(1) population.

FIGURE **14.15**
Rejection region for $\alpha = 0.05$ for a left-sided one-sample t test based on 500 samples of size $n = 10$ from an Exponential(1) population.

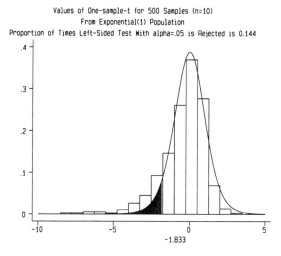

In an effort to improve our confidence in the significance level of the one-sample t test from an exponential parent distribution, we'll increase the sample size to n_1 = 40 and run the lab for both a Right-Sided and Left-Sided test. The resulting graphs are shown in Figures 14.16 and 14.17. Although the observed Type I error rates are closer to 0.05, they still leave something to be desired. We see the same pattern; that is, for the right-sided test, the actual Type I error rate is too small, and for the left-sided test, it is too large.

What about two-sided tests? Because the right-sided tests have too large a Type I error rate, and the left-tailed tests have too small a Type I error rate, will the two-tailed tests end up being pretty close? Run the lab for n_1 = 40 and Test of Two-Sided to see. The result is shown in Figure 14.18. The Type I error rate for the 500 samples is 0.082, which is larger than alpha. If you run the lab repeatedly, you will see that this is generally the case, even for a sample size as large as $n = 40$. Even if the true Type I error rate was pretty close, you still should not be comfortable with the

FIGURE **14.16**
Rejection region for $\alpha = 0.05$ for a right-sided one-sample t test based on 500 samples of size $n = 40$ from an Exponential(1) population.

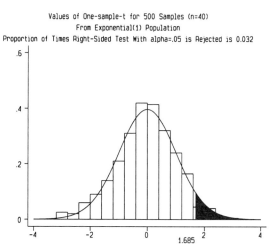

FIGURE **14.17**
Rejection region for $\alpha = 0.05$ for a left-sided one-sample t test based on 500 samples of size $n = 40$ from an Exponential(1) population.

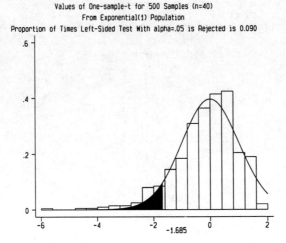

FIGURE **14.18**
Rejection regions for $\alpha = 0.05$ for a two-sided one-sample t test based on 500 samples of size $n = 40$ from an Exponential(1) population.

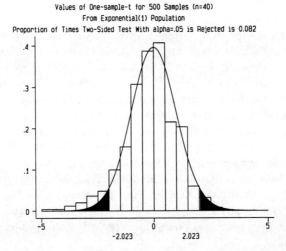

results because the shapes of the distributions are not the same. For an Exponential(1) parent, we would still be getting "lopsided" results.

14.4.3 The Two-sample t Test Statistic

The two-sample t test statistic is used for inferential procedures that compare the value of two population means. The assumptions that allow us to use a $t_{n_1+n_2-2}$ distribution as the sampling distribution of this test statistic are as follows:

1. Both parent populations are independent of each other and have distributions belonging to the normal family.
2. The variance of the two populations is the same; that is, $\sigma_1^2 = \sigma_2^2$. Note that this does not imply that we know the value of the populations' variances, only that we know that those values are equal.
3. The samples are randomly and independently selected.

If these assumptions are satisfied, then for the null hypothesis $H_0: \mu_1 - \mu_2 = D_0$, the two-sample t test statistic is

$$t = \frac{(\bar{X}_1 - \bar{X}_2) - D_0}{\sqrt{s_p^2 \left(\frac{1}{n_1} + \frac{1}{n_2} \right)}}.$$

In this expression, s_p^2 is the pooled variance and is the estimator of the common value of the population variances. The pooled variance is calculated using

$$s_p^2 = \frac{(n_1 - 1)s_1^2 + (n_2 - 1)s_2^2}{n_1 + n_2 - 2}.$$

Because the pooled variance is used in this two-sample t test statistic, this test is often referred to as the *pooled t test*.

Because the two-sample t test statistic has the same distribution (except for degrees of freedom) as the one-sample t test statistic, we will not spend much time examining the behavior for different cases. If the assumptions are satisfied, we can expect close agreement between the desired value of α and the actual proportion of tests that result in rejecting the null hypothesis.

Let n_1 = n_2 = 10, alpha = 0.05, Test be Right-Sided, Statistic be Two-sample-t, and Sample from be Normal(0,1). Run the lab. As you can see in Figure 14.19, the agreement between the two Type I error rates is remarkable. However, because the populations have a normal distribution, we are not surprised. We won't do this here, but you might want to run the lab for varying sample sizes, levels of α, and types of tests to confirm that the two error rates agree very nicely if the populations' distributions are normal.

FIGURE **14.19** Rejection region for $\alpha = 0.05$ for a right-sided two-sample t test based on 500 samples of sizes $n_1 = n_2 = 10$ from a Normal(0,1) population.

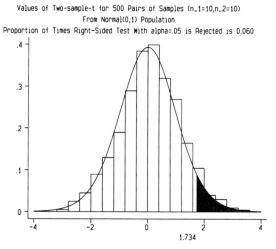

Now change Sample from to Uniform(0,1) and run the lab again. We see in Figure 14.20 that, just as in the case of the one-sample t statistic, the agreement

FIGURE 14.20
Rejection region for $\alpha = 0.05$ for a right-sided two-sample t test based on 500 samples of sizes $n_1 = n_2 = 10$ from a Uniform(0,1) population.

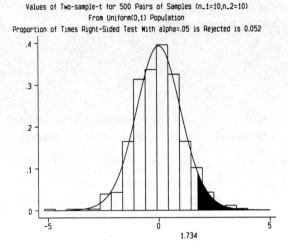

between the desired value of `alpha`, which is 0.05, and the proportion of the 500 samples that resulted in rejecting H_0, which is 0.052, is very close. Because both sample sizes are 10, based on what we have already seen for `Uniform(0,1)` populations, we would expect the two to be very close.

Figure 14.21 shows the result of running the lab for an `Exponential(1)` parent population. Notice that the proportion of samples resulting in rejecting H_0 is 0.062, which is a little too large for us to be happy with the results of using the t_{18} distribution. If both sample sizes are increased to n_1 = n_2 = 40, we see in Figure 14.22 that the agreement is much closer: The observed level of significance from the 500 samples is 0.056. Therefore, increasing the sample size does eventually assure us the $t_{n_1+n_2-2}$ distribution will be an adequate sampling distribution for purposes of statistical inference.

FIGURE 14.21
Rejection region for $\alpha = 0.05$ for a right-sided two-sample t test based on 500 samples of sizes $n_1 = n_2 = 10$ from an Exponential(1) population.

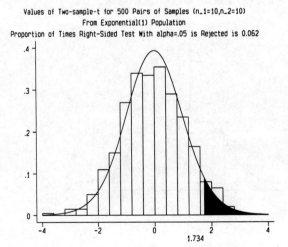

FIGURE 14.22 Rejection region for $\alpha = 0.05$ for a right-sided two-sample t test based on 500 samples of sizes $n_1 = n_2 = 40$ from an Exponential(1) population.

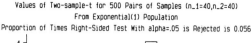
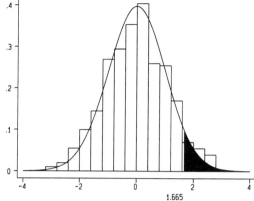

14.4.4 The χ^2 Test Statistic

The χ^2 statistic is used for making inferences on the value of the variance σ^2 of a normal population. If the null hypothesis is $H_0: \sigma^2 = \sigma_0^2$, then the χ^2 test statistic

$$\chi^2 = (n-1)\frac{s^2}{\sigma_0^2}$$

has a χ^2_{n-1} distribution.

Figure 14.23 shows a graph of the lab run for the Chi-square statistic when n_1 = 10, alpha = 0.05, Test is Right-Sided, and Sample from is Normal(0,1). Because the assumption that the population distribution is normally distributed is satisfied, the histogram of the 500 values of the χ^2 statistic closely

FIGURE 14.23 Rejection region for $\alpha = 0.05$ for a right-sided χ^2 test based on 500 samples of size $n = 10$ from a Normal(0,1) population.

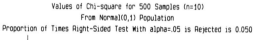
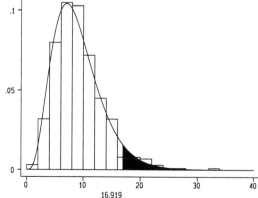

matches the χ_9^2 curve. Further, the proportion of χ^2 statistics that resulted in rejecting the null hypothesis is exactly 0.050. Again, we see that if the assumptions are satisfied, then the observed sampling distribution behaves as the theory says it will.

You may recall from the *Sampling Distributions* Lab that when the normality assumption was violated, then the Monte Carlo distribution was very far away from the theoretical χ_{n-1}^2 distribution. We now examine just how these differences affect the actual level of significance of a hypothesis test that uses the χ^2 test statistic. Change `Sample from` to `Uniform(0,1)` and run the lab again. The result should be similar to what is shown in Figure 14.24. Notice in this figure that the actual proportion of test statistics that resulted in rejecting the null hypothesis for this right-tailed test is only 0.010—astoundingly less than 0.05! Although this might seem to be a good result, when we study the power of the test, we will see that this consequence can cause difficulties when the null hypothesis is false! In Figure 14.25, we increased the sample size to n_1 = 40. In this case, the proportion of test statistics that rejected the null

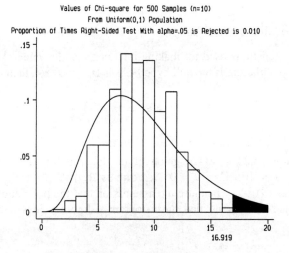

FIGURE **14.24**
Rejection region for $\alpha = 0.05$ for a right-sided χ^2 test based on 500 samples of size $n = 10$ from a Uniform(0,1) population.

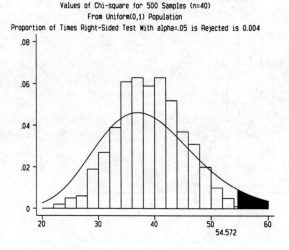

FIGURE **14.25**
Rejection region for $\alpha = 0.05$ for a right-sided χ^2 test based on 500 samples of size $n = 40$ from a Uniform(0,1) population.

hypothesis was only 0.004. Increasing the sample size actually did nothing to bring the two numbers closer together! If you run the lab for a left-sided test, you will see the same behavior, but on the left side of the distribution. That is, the actual proportion of χ^2 statistics from the 500 samples that result in rejecting H_0 will be much less than alpha, and increasing the sample size does little to rectify the situation.

In the case of the Exponential(1) distribution, we see the opposite behavior occurring. Figure 14.26 shows the result of running the lab for a sample of size n_1 = 10 for a Right-Sided test for an Exponential(1) distribution. In this case, the actual proportion of the 500 samples resulting in a Type I error is 0.110, which is astoundingly larger than 0.05. This means that if we were to use the χ_9^2 sampling distribution to make inferences on the population variance, we would actually be rejecting a true null hypothesis for around 11% of the samples, instead of only the desired 5%. For the same type of test but with a sample of size n_1 = 40, as shown in Figure 14.27, the observed proportion increases to 0.168. If you run the lab for the

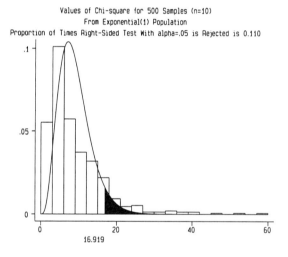

FIGURE **14.26**
Rejection region for $\alpha = 0.05$ for a right-sided χ^2 test based on 500 samples of size $n = 10$ from an Exponential(1) population.

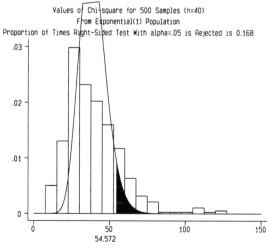

FIGURE **14.27**
Rejection region for $\alpha = 0.05$ for a right-sided χ^2 test based on 500 samples of size $n = 40$ from an Exponential(1) population.

Left-Sided test, you will see the same result. The actual proportion of samples resulting in a Type I error is larger than the desired value of alpha, and increasing the sample size does little to help. In sum, we can easily see the importance of satisfying the normality assumption for purposes of statistical inference on σ^2.

14.4.5 The F Test Statistic

For cases in which two population variances are to be compared, the F statistic is commonly used. For the null hypothesis $H_0: \sigma_1^2/\sigma_2^2 = 1$, the test statistic given by

$$F = \frac{s_1^2}{s_2^2}$$

has an F_{n_1-1, n_2-1} distribution. The assumptions that must be satisfied are that the distributions of the two independent populations belong to the normal family and the samples are chosen randomly and independently of one another.

If you have not already done so, change Statistic to F. Let n_1 = n_2 = 10, alpha = 0.05, Test be Right-Sided, and Sample from be Normal(0,1). Then run the lab. The resulting graph should be similar to the one shown in Figure 14.28. Because the normality assumption is satisfied, the sampling distribution of the 500 values of the F statistic is very close to the $F_{9,9}$ theoretical distribution. If we compare the Type I error rate of the 500 values, we see that it is 0.044, which is very close to the specified alpha = 0.05.

To examine how non-normality can affect the reliability of the inference procedure, change Sample from to Uniform(0,1) and run the lab. The result is shown in Figure 14.29. In this graph, the sampling distribution is not particularly close to the $F_{9,9}$ distribution. Nor is the observed value of the Type I error rate, which is 0.016, very close to alpha, which is 0.05. Obviously, this is problematic. Increase the sample sizes to n_1 = n_2 = 40 and run the lab again. As you can see in Figure 14.30, decreasing the sample sizes had little effect on the differences

FIGURE **14.28**
Rejection region for $\alpha = 0.05$ for a right-sided F test based on 500 samples of sizes $n_1 = n_2 = 10$ from a Normal(0,1) population.

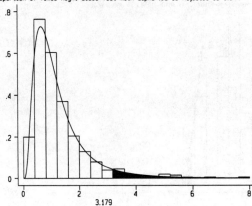

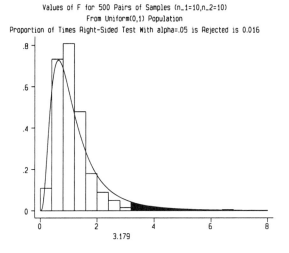

FIGURE 14.29
Rejection region for $\alpha = 0.05$ for a right-sided F test based on 500 samples of sizes $n_1 = n_2 = 10$ from a Uniform(0,1) population.

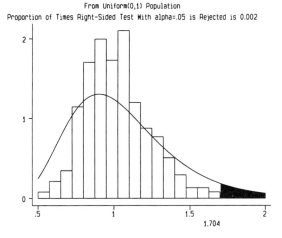

FIGURE 14.30
Rejection region for $\alpha = 0.05$ for a right-sided F test based on 500 samples of sizes $n_1 = n_2 = 40$ from a Uniform(0,1) population.

between the Monte Carlo distribution of the 500 F statistics and the theoretical $F_{39,39}$ distribution. The actual proportion of Type I errors in the 500 samples is only 0.002. If you were to run the lab again under similar circumstances for a Left-Sided or a Two-Sided test, you would see that when the parent population is Uniform(0,1), the tail areas of the true sampling distribution of the F statistic are much smaller than the tail areas of the corresponding F_{n_1-1, n_2-1} distribution.

Now change the values of n_1 and n_2 back to 10, and let the population distribution in Sample from be the Exponential(1) distribution. Run the lab for a Right-Sided test to get a graph similar to the one shown in Figure 14.31. In this graph, the proportion of F statistics to the right of the critical value $F_{0.05,9,9} = 3.179$ is 0.166. This Type I error rate is much larger than the desired alpha = 0.05. This means that if we were to use the $F_{9,9}$ sampling distribution to determine the rejection region for the test, we would actually be making a Type I error around 17% of the time, instead of just 5%. Finally, increase the sample sizes to n_1 = n_2 = 40 and

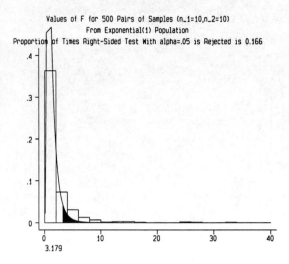

FIGURE **14.31**
Rejection region for $\alpha = 0.05$ for a right-sided F test based on 500 samples of sizes $n_1 = n_2 = 10$ from an Exponential(1) population.

run the lab again. The result is shown in Figure 14.32. Here, you can see that even with both sample sizes being somewhat large, the actual proportion of Type I errors made is 0.194, which, again, is much larger than 0.05. If you run the lab again for `Left-Sided` or `Two-Sided` tests, you will see that in both cases, the actual Type I error rate is largely compromised. Based on these results, it is not difficult to see the importance of having a normally distributed parent population when doing an F test.

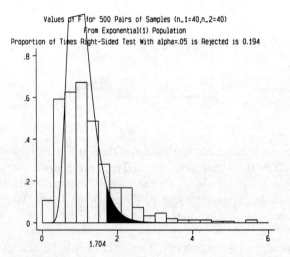

FIGURE **14.32**
Rejection region for $\alpha = 0.05$ for a right-sided F test based on 500 samples of sizes $n_1 = n_2 = 40$ from an Exponential(1) population.

14.5 Summary

The relationships witnessed in this lab can be summarized as follows:

1. In a hypothesis test with a true null hypothesis, if millions of samples of size n are drawn from the population, then α is the proportion of all of these samples

that would result in a value of the test statistic that would lead to rejecting the (true) H_0.

2. If the assumptions are satisfied for any test, the type of test will not affect the overall reliability of the results.

3. If the assumptions are satisfied for any test, the value of α will not be compromised.

4. For test statistics with symmetrical sampling distributions (like the Z or t distribution) and symmetrical parent populations (like the Normal(0,1) or Uniform(0,1)), the true Type I error rate is the same for both right-sided and left-sided hypothesis tests. For non-normal parent populations, the difference between the true and the desired Type I error rate becomes smaller as the sample size increases.

5. For the Z test statistic and an Exponential(1) parent distribution, the true Type I error rate is larger than the desired α for right-sided hypothesis tests and smaller than the desired α for left-sided hypothesis tests. This discrepancy becomes smaller as the sample size increases.

6. For the t test statistics and an Exponential(1) parent distribution, the true Type I error rate is smaller than the desired α for right-sided hypothesis tests and larger than the desired α for left-sided hypothesis tests. This discrepancy becomes smaller as the sample size increases.

7. For the χ^2 test statistic, the assumption that the population be normally distributed is crucial for reliable inference procedures. For a uniformly distributed parent, the actual Type I error rate is smaller than the desired value of α. In contrast, for the exponentially distributed parent, the observed Type I error rate is larger than α. In both cases, even relatively large sample sizes will not bring the sampling distribution of the statistic close to a χ^2_{n-1} distribution.

8. For the F test statistic, the assumption that the population be normally distributed is crucial for reliable inference procedures. For a uniformly distributed parent, the actual Type I error rate is smaller than the desired value of α. In contrast, for the exponentially distributed parent, the observed Type I error rate is larger than α. In both cases, even relatively large sample sizes will not bring the sampling distribution of the statistic close to an F_{n_1-1, n_2-1} distribution.

14.6 Lab Exercises

1. What does it mean to make a Type I error?
2. What is the difference between α and a Type I error?
3. What is the level of significance of a test, and how should it be interpreted?

14.6.1 The Z Test Statistic

In exercises 4–9, let the value of Statistic be Z.

4. Let Sample from be Normal(0,1). For a Right-Sided level alpha = 0.05 test, run the lab for the following sample sizes. Print each graph. How does changing the sample size affect the reliability of the results in this case? Repeat

this procedure for Left-Sided and Two-Sided tests, and print the graphs. Explain any differences or similarities in the resulting graphs.

(a) n_1 = 5 (b) n_1 = 15 (c) n_1 = 20 (d) n_1 = 30 (e) n_1 = 50

5 Let Sample from be Normal(0,1). For n_1 = 25, run the lab for the following values of alpha. Print each graph. How does changing the value of alpha affect the closeness of the desired level of significance and the observed level of significance? Repeat this procedure for Left-Sided and Two-Sided tests, and print the graphs. Explain any differences or similarities in the results.

(a) alpha = 0.01 (b) alpha = 0.025 (c) alpha = 0.10

6 Let Sample from be Uniform(0,1). For a Right-Sided level alpha = 0.05 test, run the lab for the following sample sizes. Print each graph. How does changing the sample size affect the reliability of the results in this case? Compare these results to those in exercise 4. Repeat this procedure for Left-Sided and Two-Sided tests, and print the graphs. Explain any differences or similarities in the resulting graphs. Compare these results to those in exercise 4.

(a) n_1 = 5 (b) n_1 = 15 (c) n_1 = 20 (d) n_1 = 30 (e) n_1 = 50

7 Let Sample from be Uniform(0,1). For n_1 = 25, run the lab for the following values of alpha. Print each graph. How does changing the value of alpha affect the closeness of the desired level of significance and the observed level of significance? Compare these results to those in exercise 5. Repeat this procedure for Left-Sided and Two-Sided tests and print the graphs. Explain any differences or similarities in the results. Compare these results to those in exercise 5.

(a) alpha = 0.01 (b) alpha = 0.025 (c) alpha = 0.10

8 Let Sample from be Exponential(1). For a Right-Sided level alpha = 0.05 test, run the lab for the following sample sizes. Print each graph. How does changing the sample size affect the reliability of the results in this case? Compare these results to those in exercises 4 and 6. Repeat this procedure for Left-Sided and Two-Sided tests, and print the graphs. Explain any differences or similarities in the resulting graphs. Compare these results to those in exercises 4 and 6.

(a) n_1 = 5 (b) n_1 = 15 (c) n_1 = 20 (d) n_1 = 30 (e) n_1 = 50

9 Let Sample from be Exponential(1). For n_1 = 25, run the lab for the following values of alpha. Print each graph. How does changing the value of alpha affect the closeness of the desired level of significance and the observed level of significance? Compare these results to those in exercises 5 and 7. Repeat this procedure for Left-Sided and Two-Sided tests, and print the graphs. Explain any differences or similarities in the results. Compare these results to those in exercises 5 and 7.

(a) alpha = 0.01 (b) alpha = 0.025 (c) alpha = 0.10

14.6.2 The One-sample t Test Statistic

In exercises 10–15, let the value of Statistic be One-sample-t.

10. Let Sample from be Normal(0,1). For a Right-Sided level alpha = 0.05 test, run the lab for the following sample sizes. Print each graph. How does changing the sample size affect the reliability of the results in this case? Repeat this procedure for Left-Sided and Two-Sided tests, and print the graphs. Explain any differences or similarities in the resulting graphs.

    (a) n_1 = 5   (b) n_1 = 15   (c) n_1 = 20   (d) n_1 = 30   (e) n_1 = 50

11. Let Sample from be Normal(0,1). For n_1 = 25, run the lab for the following values of alpha. Print each graph. How does changing the value of alpha affect the closeness of the desired level of significance and the observed level of significance? Repeat this procedure for Left-Sided and Two-Sided tests, and print the graphs. Explain any differences or similarities in the results.

    (a) alpha = 0.01   (b) alpha = 0.025   (c) alpha = 0.10

12. Let Sample from be Uniform(0,1). For a Right-Sided level alpha = 0.05 test, run the lab for the following sample sizes. Print each graph. How does changing the sample size affect the reliability of the results in this case? Compare these results to those in exercise 10. Repeat this procedure for Left-Sided and Two-Sided tests, and print the graphs. Explain any differences or similarities in the resulting graphs. Compare these results to those in exercise 10.

    (a) n_1 = 5   (b) n_1 = 15   (c) n_1 = 20   (d) n_1 = 30   (e) n_1 = 50

13. Let Sample from be Uniform(0,1). For n_1 = 25, run the lab for the following values of alpha. Print each graph. How does changing the value of alpha affect the closeness of the desired level of significance and the observed level of significance? Compare these results to those in exercise 11. Repeat this procedure for Left-Sided and Two-Sided tests, and print the graphs. Explain any differences or similarities in the results. Compare these results to those in exercise 11.

    (a) alpha = 0.01   (b) alpha = 0.025   (c) alpha = 0.10

14. Let Sample from be Exponential(1). For a Right-Sided level alpha = 0.05 test, run the lab for the following sample sizes. Print each graph. How does changing the sample size affect the reliability of the results in this case? Compare these results to those in exercises 10 and 12. Repeat this procedure for Left-Sided and Two-Sided tests, and print the graphs. Explain any differences or similarities in the resulting graphs. Compare these results to those in exercises 10 and 12.

    (a) n_1 = 5   (b) n_1 = 15   (c) n_1 = 20   (d) n_1 = 30   (e) n_1 = 50

15. Let Sample from be Exponential(1). For n_1 = 25, run the lab for the following values of alpha. Print each graph. How does changing the value of alpha affect the closeness of the desired level of significance and the observed level of significance? Compare these results to those in exercises 11 and 13. Repeat

this procedure for `Left-Sided` and `Two-Sided` tests, and print the graphs. Explain any differences or similarities in the results. Compare these results to those in exercises 11 and 13.

(a) `alpha = 0.01`   (b) `alpha = 0.025`   (c) `alpha = 0.10`

14.6.3 The Two-sample t Test Statistic

In exercises 16–21, let the value of `Statistic` be `Two-sample-t`.

16 Let `Sample from` be `Normal(0,1)`. For a `Right-Sided` level `alpha = 0.05` test, run the lab for the following sample sizes. Print each graph. How does changing the sample size affect the reliability of the results in this case? Repeat this procedure for `Left-Sided` and `Two-Sided` tests, print the graphs. Explain any differences or similarities in the resulting graphs.

(a) $n_1 = n_2 = 5$ (b) $n_1 = n_2 = 15$ (c) $n_1 = n_2 = 20$
(d) $n_1 = n_2 = 30$ (e) $n_1 = n_2 = 50$

17 Let `Sample from` be `Normal(0,1)`. For $n_1 = n_2 = 25$, run the lab for the following values of `alpha`. Print each graph. How does changing the value of `alpha` affect the closeness of the desired level of significance and the observed level of significance? Repeat this procedure for `Left-Sided` and `Two-Sided` tests, and print the graphs. Explain any differences or similarities in the results.

(a) `alpha = 0.01`   (b) `alpha = 0.025`   (c) `alpha = 0.10`

18 Let `Sample from` be `Uniform(0,1)`. For a `Right-Sided` level `alpha = 0.05` test, run the lab for the following sample sizes. Print each graph. How does changing the sample size affect the reliability of the results in this case? Compare these results to those in exercise 16. Repeat this procedure for `Left-Sided` and `Two-Sided` tests, and print the graphs. Explain any differences or similarities in the resulting graphs. Compare these results to those in exercise 16.

(a) $n_1 = n_2 = 5$ (b) $n_1 = n_2 = 15$ (c) $n_1 = n_2 = 20$
(d) $n_1 = n_2 = 30$ (e) $n_1 = n_2 = 50$

19 Let `Sample from` be `Uniform(0,1)`. For $n_1 = n_2 = 25$, run the lab for the following values of `alpha`. Print each graph. How does changing the value of `alpha` affect the closeness of the desired level of significance and the observed level of significance? Compare these results to those in exercise 17. Repeat this procedure for `Left-Sided` and `Two-Sided` tests, and print the graphs. Explain any differences or similarities in the results. Compare these results to those in exercise 17.

(a) `alpha = 0.01`   (b) `alpha = 0.025`   (c) `alpha = 0.10`

20 Let `Sample from` be `Exponential(1)`. For a `Right-Sided` level `alpha = 0.05` test, run the lab for the following sample sizes. Print each graph. How does changing the sample size affect the reliability of the results in this case? Compare these results to those in exercises 16 and 18. Repeat this procedure for `Left-Sided` and `Two-Sided` tests, and print the graphs. Explain any

differences or similarities in the resulting graphs. Compare these results to those in exercises 16 and 18.

(a) n_1 = n_2 = 5 (b) n_1 = n_2 = 15 (c) n_1 = n_2 = 20
(d) n_1 = n_2 = 30 (e) n_1 = n_2 = 50

21 Let Sample from be Exponential(1). For n_1 = n_2 = 25, run the lab for the following values of alpha. Print each graph. How does changing the value of alpha affect the closeness of the desired level of significance and the observed level of significance? Compare these results to those in exercises 17 and 19. Repeat this procedure for Left-Sided and Two-Sided tests, and print the graphs. Explain any differences or similarities in the results. Compare these results to those in exercises 17 and 19.

(a) alpha = 0.01 (b) alpha = 0.025 (c) alpha = 0.10

14.6.4 The χ^2 Test Statistic

In exercises 22–27, let the value of Statistic be Chi-square.

22 Let Sample from be Normal(0,1). For a Right-Sided level alpha = 0.05 test, run the lab for the following sample sizes. Print each graph. How does changing the sample size affect the reliability of the results in this case? Repeat this procedure for Left-Sided and Two-Sided tests, and print the graphs. Explain any differences or similarities in the resulting graphs.

(a) n_1 = 5 (b) n_1 = 15 (c) n_1 = 20 (d) n_1 = 30 (e) n_1 = 50

23 Let Sample from be Normal(0,1). For n_1 = 25, run the lab for the following values of alpha. Print each graph. How does changing the value of alpha affect the closeness of the desired level of significance and the observed level of significance? Repeat this procedure for Left-Sided and Two-Sided tests, and print the graphs. Explain any differences or similarities in the results.

(a) alpha = 0.01 (b) alpha = 0.025 (c) alpha = 0.10

24 Let Sample from be Uniform(0,1). For a Right-Sided level alpha = 0.05 test, run the lab for the following sample sizes. Print each graph. How does changing the sample size affect the reliability of the results in this case? Compare these results to those in exercise 22. Repeat this procedure for Left-Sided and Two-Sided tests, and print the graphs. Explain any differences or similarities in the resulting graphs. Compare these results to those in exercise 22.

(a) n_1 = 5 (b) n_1 = 15 (c) n_1 = 20 (d) n_1 = 30 (e) n_1 = 50

25 Let Sample from be Uniform(0,1). For n_1 = 25, run the lab for the following values of alpha. Print each graph. How does changing the value of alpha affect the closeness of the desired level of significance and the observed level of significance? Compare these results to those in exercise 23. Repeat this procedure for Left-Sided and Two-Sided tests, and print the graphs. Explain any differences or similarities in the results. Compare these results to those in exercise 23.

(a) alpha = 0.01 (b) alpha = 0.025 (c) alpha = 0.10

26 Let `Sample from` be `Exponential(1)`. For a `Right-Sided` level `alpha = 0.05` test, run the lab for the following sample sizes. Print each graph. How does changing the sample size affect the reliability of the results in this case? Compare these results to those in exercises 22 and 24. Repeat this procedure for `Left-Sided` and `Two-Sided` tests, and print the graphs. Explain any differences or similarities in the resulting graphs. Compare these results to those in exercises 22 and 24.

(a) $n\_1 = 5$ (b) $n\_1 = 15$ (c) $n\_1 = 20$ (d) $n\_1 = 30$ (e) $n\_1 = 50$

27 Let `Sample from` be `Exponential(1)`. For $n\_1 = 25$, run the lab for the following values of `alpha`. Print each graph. How does changing the value of `alpha` affect the closeness of the desired level of significance and the observed level of significance? Compare these results to those in exercises 23 and 25. Repeat this procedure for `Left-Sided` and `Two-Sided` tests, and print the graphs. Explain any differences or similarities in the results. Compare these results to those in exercises 23 and 25.

(a) `alpha = 0.01`   (b) `alpha = 0.025`   (c) `alpha = 0.10`

14.6.5 The F Statistic

In exercises 28–33, let the value of `Statistic` be F.

28 Let `Sample from` be `Normal(0,1)`. For a `Right-Sided` level `alpha = 0.05` test, run the lab for the following sample sizes. Print each graph. How does changing the sample size affect the reliability of the results in this case? Repeat this procedure for `Left-Sided` and `Two-Sided` tests, and print the graphs. Explain any differences or similarities in the resulting graphs.

(a) $n\_1 = n\_2 = 5$ (b) $n\_1 = n\_2 = 15$ (c) $n\_1 = n\_2 = 20$
(d) $n\_1 = n\_2 = 30$ (e) $n\_1 = n\_2 = 50$

29 Let `Sample from` be `Normal(0,1)`. For $n\_1 = n\_2 = 25$, run the lab for the following values of `alpha`. Print each graph. How does changing the value of `alpha` affect the closeness of the desired level of significance and the observed level of significance? Repeat this procedure for `Left-Sided` and `Two-Sided` tests, and print the graphs. Explain any differences or similarities in the results.

(a) `alpha = 0.01`   (b) `alpha = 0.025`   (c) `alpha = 0.10`

30 Let `Sample from` be `Uniform(0,1)`. For a `Right-Sided` level `alpha = 0.05` test, run the lab for the following sample sizes. Print each graph. How does changing the sample size affect the reliability of the results in this case? Compare these results to those in exercise 28. Repeat this procedure for `Left-Sided` and `Two-Sided` tests, and print the graphs. Explain any differences or similarities in the resulting graphs. Compare these results to those in exercise 28.

(a) $n\_1 = n\_2 = 5$ (b) $n\_1 = n\_2 = 15$ (c) $n\_1 = n\_2 = 20$
(d) $n\_1 = n\_2 = 30$ (e) $n\_1 = n\_2 = 50$

31 Let `Sample from` be `Uniform(0,1)`. For $n\_1 = n\_2 = 25$, run the lab for the following values of `alpha`. Print each graph. How does changing the value of `alpha` affect the closeness of the desired level of significance and the observed

level of significance? Compare these results to those in exercise 29. Repeat this procedure for Left-Sided and Two-Sided tests, and print the graphs. Explain any differences or similarities in the results. Compare these results to those in exercise 29.

(a) alpha = 0.01 (b) alpha = 0.025 (c) alpha = 0.10

32 Let Sample from be Exponential(1). For a Right-Sided level alpha = 0.05 test, run the lab for the following sample sizes. Print each graph. How does changing the sample size affect the reliability of the results in this case? Compare these results to those in exercises 28 and 30. Repeat this procedure for Left-Sided and Two-Sided tests, and print the graphs. Explain any differences or similarities in the resulting graphs. Compare these results to those in exercises 28 and 30.

(a) $n_1 = n_2 = 5$ (b) $n_1 = n_2 = 15$ (c) $n_1 = n_2 = 20$
(d) $n_1 = n_2 = 30$ (e) $n_1 = n_2 = 50$

33 Let Sample from be Exponential(1). For $n_1 = n_2 = 25$, run the lab for the following values of alpha. Print each graph. How does changing the value of alpha affect the closeness of the desired level of significance and the observed level of significance? Compare these results to those in exercises 29 and 31. Repeat this procedure for Left-Sided and Two-Sided tests, and print the graphs. Explain any differences or similarities in the results. Compare these results to those in exercises 29 and 31.

(a) alpha = 0.01 (b) alpha = 0.025 (c) alpha = 0.10

15

Calculating Tests of Hypotheses

Chapters 13 and 14 were concerned with the ideas of testing hypotheses. Now we do some examples with real data.

15.1 Introduction

This lab brings together all of the one- and two-sample tests of hypotheses for means, variances, and proportions typically covered in introductory statistics courses. See the *Level of Significance of a Test* Lab for a review of tests of hypotheses.

15.2 Objectives

This lab is primarily computational. It makes it easy to perform tests and to interpret the results.

15.3 Description

Given summary statistics (sample means, variances, proportions) for a test-of-hypothesis problem, the lab is used as follows. To begin the lab, select `Calculating Tests of Hypotheses` from the `Labs` menu. The dialog box shown in Figure 15.1 appears; this box contains a number of items:

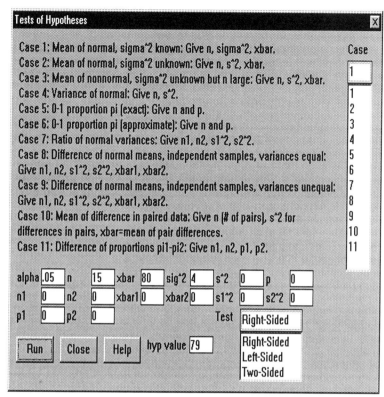

FIGURE 15.1
Dialog box for the *Calculating Tests of Hypotheses* Lab.

1. A list of 11 test-of-hypothesis *cases* usually taught in introductory statistics courses. At the end of this chapter are the formulas needed for the cases. For each case, the dialog box specifies what input is required.

2. A box labeled Case, where you select the number of the desired case. The initial entry is 1.

3. A box labeled Test, where you select whether the test is Left-Sided, Right-Sided, or Two-Sided. The initial value is Right-Sided.

4. A set of boxes where you enter required input, including sample size(s), mean(s), variance(s), proportion(s), and two other values:

    (a) alpha: This is the level of significance to be used in the test. The initial value is 0.05.

    (b) hyp value: This is the hypothesized value of the parameter being tested.

5. The usual three buttons:

    (a) Run: After selecting Case and Test, and filling in hyp value, alpha, and any other information required for the test, click on Run to execute the lab. The numerical results of the lab are displayed in the results window, including a description of the test, the input supplied, the calculated value of the test statistic, whether the null hypothesis is rejected, and the p value of the test. This same information is displayed at the top of a graphics window, as is a plot of the appropriate curve (Z, t, χ^2, or F) for the test. The value of the test statistic is marked on the curve in yellow and the critical value(s) is marked in red so that you can see whether the calculated value falls in the rejection region of the test. Finally, the area under the curve corresponding to the p value of the test is shaded in yellow.

(b) `Close`: Clicking on this button closes the lab and returns you to the Stat-Concepts menus.

(c) `Help`: Clicking on this button opens a help window containing information about the lab.

When the lab begins, all of the information required for the first example in our guided tour has already been entered into the dialog box. Thus, clicking on `Run` shows the analysis for that example.

15.4 Guided Tour of the Lab

In this section, we consider three examples of calculating tests of hypotheses, the last of which has two hypothesis tests. For clarity we subscript our test statistics with a C.

15.4.1 Example 1

In a study of the keyboard height that typists prefer, a sample of 15 trained typists was selected and their preferred heights determined. The resulting average height was 80 cm. Assuming that the population of all typists' preferred heights is normally distributed with standard deviation 2 cm, is this enough evidence to conclude that the average preferred height is more than 79 cm?

Analysis of Example 1

Because $Z_C = 1.936 > 1.645 = Z_{0.05}$, we reject $H_0: \mu \leq 79$ at the $\alpha = 0.05$ level of significance and conclude that there is enough evidence to say that the average preferred height is more than 79 cm. The graph in Figure 15.2 makes this very clear, because the calculated Z is to the right of the critical value. Note that the p value is 0.026 (the shaded area to the right of the calculated value of Z), so this is strong, but not overwhelming, evidence against H_0 (for example, we would not reject H_0 for $\alpha = 0.025$).

Results Window for Example 1

```
Right-Sided Test for Mean of normal, sigma known:
n = 15, sigma^2 = 4, xbar = 80
alpha = .05
Hypothesized value = 79

Z_calc = 1.9364917
Critical value: 1.6448536
Reject H_0
p-value = .02640376
```

FIGURE 15.2
Graphics window for example 1.

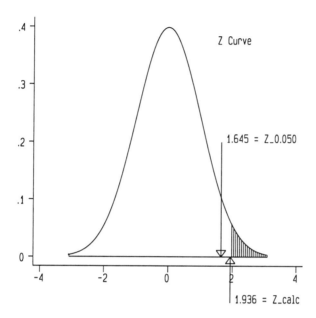

15.4.2 Example 2

A ball bearing manufacturer claims that it has developed a new process that leads to a reduction in the variation in the diameters of the ball bearings its plant produces. The current variance is 10. A random sample of 50 ball bearings produced using the new process is obtained, for which $s = 3$. Does this justify the claim?

The Analysis of Example 2

Here, we use a left-sided test because the claim is that $\sigma^2 < 10$. Also, because the parameter of interest is σ^2, the χ^2 statistic is used. The calculated value is $\chi_C^2 = 44.1$, which is not less than the critical value $\chi_{0.95,49}^2 = 33.930$, so we can't reject H_0. Note that now the p value is the area to the *left* of χ_C^2 (because we're doing a left-sided test) and is equal to 0.328 (see Figure 15.3), which means there is, in fact, little evidence against H_0.

Results Window for Example 2

```
Left-Sided Test for Variance of normal:
n = 50, s^2 = 9
alpha = .05
Hypothesized value = 10

chisquare_calc= 44.1
Critical value: 33.929901
Fail to reject H_0
p-value = .32833193
```

FIGURE 15.3
Graphics window for example 2.

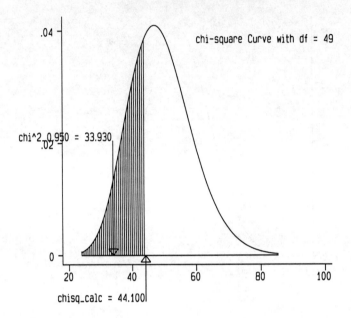

15.4.3 Example 3

To compare two teaching methods, a school found 45 students with similar backgrounds and randomly selected 25 students to receive the first method and 20 students to receive the second. At the end of the school year, the same test was administered to both groups and the results recorded as follows: $\bar{X}_1 = 75$, $\bar{X}_2 = 79$, $s_1^2 = 9$, and $s_2^2 = 16$.

(a) Assuming that the populations of test scores are normally distributed, is it reasonable to conclude that the variability in the two populations is the same?

(b) Based on your result in part (a), test the hypothesis that the two testing methods lead to the same average test score.

Analysis of Example 3(a)

Now we have the two-sided version of Case 7. That is, if we let σ_1^2 and σ_2^2 be the variances of the two populations, then we have $H_0 : \sigma_1^2 = \sigma_2^2$, with the alternative hypothesis being that the two variances are unequal. There are now two critical values, namely, $F_{0.025, 24, 19} = 2.452$ and $F_{0.975, 24, 19} = 0.426$, and the calculated value is $F_C = 0.562$, which is not more extreme (farther from 1) than either critical value. Thus, we don't reject H_0, but rather conclude that it is reasonable to assume the two variances are the same. Note that the p value is 0.183, which is the sum of the two shaded areas in Figure 15.4.

FIGURE 15.4
Graphics window for example 3(a).

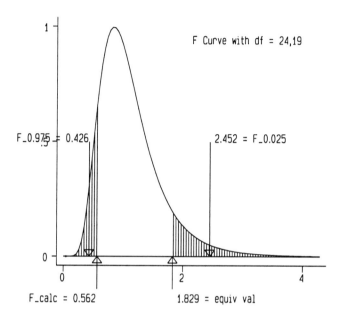

Results Window for Example 3(a)

```
Two-Sided Test for Ratio of normal variances:
n1 = 25, n2 = 20, s1^2 = 9, s2^2 = 16
alpha = .05
Hypothesized value = 1

F_calc = .5625
Critical values: .42641163 , 2.4523139
Fail to reject H_0
p-value = .18282865
```

Analysis of Example 3(b)

Because we concluded that the variances were not different in the previous example, we now have Case 8, and it is two-sided because the question only asks if the two average test scores are the same. The graph in Figure 15.5 shows that for these sample sizes and variances, the means $\bar{X}_1 = 75$ and $\bar{X}_2 = 79$ are significantly different. The test statistic is $t_C = -3.834$, and the critical values are ± 2.017. The location of the arrows on the graph emphasizes how extreme the difference in means is, as does the fact that to three decimal places, the p value is 0 (in the results window, you can see that it is actually 0.0004).

FIGURE 15.5
Graphics window for example 3(b).

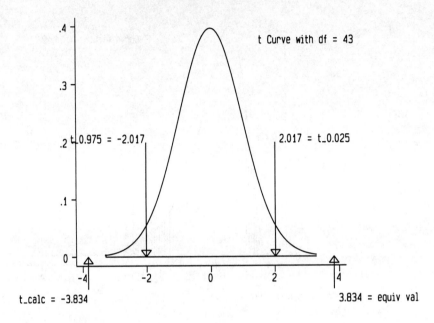

Two-Sided Test for Difference of normal means, ind samples, vars equal:
n's: 25,20 xbars: 75,79 var's: 9,16 alpha = .05 hyp val = 0
p-value = shaded area = 0.000

Results Window for Example 3(b)

```
Two-Sided Test for Difference of normal means,
  ind samples, vars equal:
n1 = 25, n2 = 20, xbar1 = 75, xbar2 = 79
s1^2 = 9, s2^2 = 16, sp^2 = 12.093023, df = 43
alpha = .05
Hypothesized value = 0

t_calc = -3.8341694
Critical values: -2.0166922 , 2.0166922
Reject H_0
p-value = .0004064
```

15.5 Summary

This lab makes it easy to perform hypothesis tests for the situations usually covered in introductory statistics courses. The graphs produced by the lab make it clear how tests work (checking to see if the test statistic is in the rejection region) and what a *p* value represents (the area under a curve for extreme values of the test statistic).

15.6 Confidence Interval and Tests of Hypotheses Formulas

Test Statistic

| H_0 | H_a | Rej H_0 if | p value |

Confidence Interval

Case 1: Normal mean μ (σ^2 known) (`1 sample normal`)

$$Z_C = \frac{\bar{X} - \mu_0}{\sigma/\sqrt{n}}$$

$\mu \leq \mu_0$	$\mu > \mu_0$	$Z_C > Z_\alpha$	$\Pr(Z > Z_C)$
$\mu \geq \mu_0$	$\mu < \mu_0$	$Z_C < -Z_\alpha$	$\Pr(Z < Z_C)$
$\mu = \mu_0$	$\mu \neq \mu_0$	$|Z_C| > Z_{\alpha/2}$	$2\Pr(Z > |Z_C|)$

$$\bar{X} \pm Z_{\alpha/2} \frac{\sigma}{\sqrt{n}}$$

Case 2: Normal mean μ (σ^2 unknown) (`1 sample t`)

$$t_C = \frac{\bar{X} - \mu_0}{s/\sqrt{n}}$$

$\mu \leq \mu_0$	$\mu > \mu_0$	$t_C > t_{\alpha,\nu}$	$\Pr(t > t_C)$
$\mu \geq \mu_0$	$\mu < \mu_0$	$t_C < -t_{\alpha,\nu}$	$\Pr(t < t_C)$
$\mu = \mu_0$	$\mu \neq \mu_0$	$|t_C| > t_{\alpha/2,\nu}$	$2\Pr(t > |t_C|)$

$$\bar{X} \pm t_{\alpha/2,\nu} \frac{s}{\sqrt{n}}, \quad \nu = n - 1$$

Case 3: Nonnormal mean μ (σ^2 unknown, n large) (`1 sample t`): The formulas are the same as for Case 2.

Case 4: Normal variance σ^2 (1 sample variance)

$$\chi_C^2 = \frac{(n-1)s^2}{\sigma_0^2}$$

$\sigma^2 \leq \sigma_0^2$	$\sigma^2 > \sigma_0^2$	$\chi_C^2 > \chi_{\alpha,\nu}^2$	$\Pr(\chi^2 > \chi_C^2)$
$\sigma^2 \geq \sigma_0^2$	$\sigma^2 < \sigma_0^2$	$\chi_C^2 < \chi_{1-\alpha,\nu}^2$	$\Pr(\chi^2 < \chi_C^2)$
$\sigma^2 = \sigma_0^2$	$\sigma^2 \neq \sigma_0^2$	$\chi_C^2 > \chi_{\alpha/2,\nu}^2$	$2\min(\Pr(\chi^2 > \chi_C^2),$
		or $\chi_C^2 < \chi_{1-\alpha/2,\nu}^2$	$1 - \Pr(\chi^2 > \chi_C^2))$

$$\left(\frac{(n-1)s^2}{\chi_{\alpha/2,\nu}^2}, \frac{(n-1)s^2}{\chi_{1-\alpha/2,\nu}^2}\right), \quad \nu = n-1$$

Case 5: Proportion π (exact CI for small n)

$$\left(\frac{XF_{1-\alpha/2,\nu_1,\nu_2}}{(n-X+1) + XF_{1-\alpha/2,\nu_1,\nu_2}}, \frac{(X+1)F_{\alpha/2,\nu_3,\nu_4}}{(n-X) + (X+1)F_{\alpha/2,\nu_3,\nu_4}}\right),$$

$$X = np, \quad \nu_1 = 2X, \quad \nu_2 = 2(n-X+1), \quad \nu_3 = 2(X+1), \quad \nu_4 = 2(n-X)$$

Case 6: Proportion π $(\min(n\pi, n(1-\pi)) \geq 5)$ (1 sample proportion)

$$Z_C = \frac{p - \pi_0}{\sqrt{\pi_0(1-\pi_0)/n}}$$

$\pi \leq \pi_0$	$\pi > \pi_0$	$Z_C > Z_\alpha$	$\Pr(Z > Z_C)$
$\pi \geq \pi_0$	$\pi < \pi_0$	$Z_C < -Z_\alpha$	$\Pr(Z < Z_C)$
$\pi = \pi_0$	$\pi \neq \pi_0$	$|Z_C| > Z_{\alpha/2}$	$2\Pr(Z > |Z_C|)$

$$p \pm Z_{\alpha/2}\sqrt{p(1-p)/n}$$

Case 7: Ratio of normal variances (independent samples) (2 sample variances)

$$F_C = \frac{s_1^2/s_2^2}{\delta}$$

$\sigma_1^2/\sigma_2^2 \leq \delta$	$\sigma_1^2/\sigma_2^2 > \delta$	$F_C > F_{\alpha,\nu_1,\nu_2}$	$\Pr(F > F_C)$
$\sigma_1^2/\sigma_2^2 \geq \delta$	$\sigma_1^2/\sigma_2^2 < \delta$	$F_C < F_{1-\alpha,\nu_1,\nu_2}$	$\Pr(F < F_C)$
$\sigma_1^2/\sigma_2^2 = \delta$	$\sigma_1^2/\sigma_2^2 \neq \delta$	$F_C > F_{\alpha/2,\nu_1,\nu_2}$	$2\min(\Pr(F > F_C),$
		or $F_C < F_{1-\alpha/2,\nu_1,\nu_2}$	$1 - \Pr(F > F_C))$

$$\left(\frac{s_1^2/s_2^2}{F_{\alpha/2,\nu_1,\nu_2}}, \frac{s_1^2/s_2^2}{F_{1-\alpha/2,\nu_1,\nu_2}}\right), \quad \nu_1 = n_1 - 1, \quad \nu_2 = n_2 - 1$$

15.6 Confidence Interval and Tests of Hypotheses Formulas

Case 8: Difference of two normal means (σ_1^2, σ_2^2 unknown, but equal, independent samples) (2 sample t)

$$t_C = \frac{(\bar{X}_1 - \bar{X}_2) - \delta}{\sqrt{s_P^2(1/n_1 + 1/n_2)}}$$

$\mu_1 - \mu_2 \leq \delta$	$\mu_1 - \mu_2 > \delta$	$t_C > t_{\alpha,\nu}$	$\Pr(t > t_C)$
$\mu_1 - \mu_2 \geq \delta$	$\mu_1 - \mu_2 < \delta$	$t_C < -t_{\alpha,\nu}$	$\Pr(t < t_C)$
$\mu_1 - \mu_2 = \delta$	$\mu_1 - \mu_2 \neq \delta$	$\|t_C\| > t_{\alpha/2,\nu}$	$2\Pr(t > \|t_C\|)$

$$(\bar{X}_1 - \bar{X}_2) \pm t_{\alpha/2,\nu}\sqrt{s_P^2(1/n_1 + 1/n_2)}, \qquad s_P^2 = \frac{(n_1 - 1)s_1^2 + (n_2 - 1)s_2^2}{n_1 + n_2 - 2},$$
$$\nu = n_1 + n_2 - 2$$

Case 9: Difference of two normal means (σ_1^2, σ_2^2 unknown, unequal, independent samples) (2 sample t)

$$t_C = \frac{(\bar{X}_1 - \bar{X}_2) - \delta}{\sqrt{s_1^2/n_1 + s_2^2/n_2}}$$

$\mu_1 - \mu_2 \leq \delta$	$\mu_1 - \mu_2 > \delta$	$t_C > t_{\alpha,\nu}$	$\Pr(t > t_C)$
$\mu_1 - \mu_2 \geq \delta$	$\mu_1 - \mu_2 < \delta$	$t_C < -t_{\alpha,\nu}$	$\Pr(t < t_C)$
$\mu_1 - \mu_2 = \delta$	$\mu_1 - \mu_2 \neq \delta$	$t_C > \|t_{\alpha/2,\nu}\|$	$2\Pr(t > \|t_C\|)$

$$(\bar{X}_1 - \bar{X}_2) \pm t_{\alpha/2,\nu}\sqrt{s_1^2/n_1 + s_2^2/n_2}, \qquad \nu = \frac{\left(s_1^2/n_1 + s_2^2/n_2\right)^2}{\dfrac{\left(s_1^2/n_1\right)^2}{n_1 - 1} + \dfrac{\left(s_2^2/n_2\right)^2}{n_2 - 1}}$$

Case 10: Mean of differences in normal paired data (σ^2 unknown) (paired t)

$$t_C = \frac{\bar{d} - \mu_0}{s_d/\sqrt{n}}$$

$\mu_d \leq \mu_0$	$\mu_d > \mu_0$	$t_C > t_{\alpha,\nu}$	$\Pr(t > t_C)$
$\mu_d \geq \mu_0$	$\mu_d < \mu_0$	$t_C < -t_{\alpha,\nu}$	$\Pr(t < t_C)$
$\mu_d = \mu_0$	$\mu_d \neq \mu_0$	$\|t_C\| > t_{\alpha/2,\nu}$	$2\Pr(t > \|t_C\|)$

$$\bar{d} \pm t_{\alpha/2,\nu}\frac{s_d}{\sqrt{n}}, \qquad \nu = n - 1, \qquad \bar{d} \text{ and } s_d \text{ are mean and standard deviation of differences in pairs.}$$

Case 11: Difference $\pi_1 - \pi_2$ of proportions (independent samples, $\min(n_1\pi_1, n_1(1-\pi_1)) \geq 5$, $\min(n_2\pi_2, n_2(1-\pi_2)) \geq 5$) (2 sample proportions)

$$Z_C = \frac{p_1 - p_2}{\sqrt{p(1-p)(1/n_1 + 1/n_2)}}$$

$\pi_1 - \pi_2 \leq 0$	$\pi_1 - \pi_2 > 0$	$Z_C > Z_\alpha$	$\Pr(Z > Z_C)$
$\pi_1 - \pi_2 \geq 0$	$\pi_1 - \pi_2 < 0$	$Z_C < -Z_\alpha$	$\Pr(Z < Z_C)$
$\pi_1 - \pi_2 = 0$	$\pi_1 - \pi_2 \neq 0$	$|Z_C| > Z_{\alpha/2}$	$2\Pr(Z > |Z_C|)$

$$(p_1 - p_2) \pm Z_{\alpha/2}\sqrt{\frac{p_1(1-p_1)}{n_1} + \frac{p_2(1-p_2)}{n_2}}, \qquad p = \frac{n_1 p_1 + n_2 p_2}{n_1 + n_2}$$

16

Power of a Test

Hypothesis testing is a statistical tool that is used in many different fields to make inferences on some characteristic of a population or to compare two populations. From sociology to biology, from medicine to production, hypothesis tests are used every day in the decision-making process.

16.1 Introduction

In the *Sampling Distributions* Lab, we studied five statistics and their sampling distributions under various conditions. We also saw how the sampling distributions were affected if those conditions were not met. Then, in the *Level of Significance of a Test* Lab, we studied the proportion of times a true null hypothesis was rejected for those five test statistics and how violating the assumptions affected that proportion. In this lab, we will examine the proportion of samples that will result in rejecting a false null hypothesis, also known as the *power of the test*.

16.1.1 Some Basic Ideas

The *Level of Significance of a Test* Lab introduced the structure of hypothesis tests. Each hypothesis test has the same basic elements.

1   A null hypothesis H_0 and an alternative hypothesis H_a.

2 A test statistic. In this lab, we will study the Z, one-sample t, two-sample t, χ^2, and F test statistics. These are the same five statistics studied in the *Sampling Distributions* and *Level of Significance of a Test* Labs.

3 A rejection rule based on the sampling distribution of the test statistic under the null hypothesis. This rule is typically in the form of a rejection region or a p value.

4 A decision in terms of the null hypothesis, stated as either "fail to reject H_0," or "reject H_0."

Interpreting Power

Suppose the null hypothesis is false, but our sample results in a value of the test statistic that leads us to fail to reject that null hypothesis. Then we are saying that the null hypothesis could be true when, in fact, it is not. This particular error is known as a *Type II error*. The proportion of all samples of size n that would result in a Type II error is typically denoted by β. Conversely, suppose our sample results in a value of the test statistic that leads us to reject this false null hypothesis. Then we have made a correct decision. The proportion of all samples of size n that will result in this type of a correct decision is the power of the test and is equal to $1 - \beta$.

In practice, when a test of hypothesis is performed, we do not know whether the null hypothesis is true. But if we know the sampling distribution of the test statistic, we can use that distribution to make statements about how confident we are that our conclusion is correct. The conclusion of a hypothesis test determines what type of error we could be making. If we reject H_0, we might be making a Type I error; if we fail to reject H_0, we might be making a Type II error. In this lab, we will be studying the chance β that a Type II error was made by examining the power of the test, $1 - \beta$. Because the power of the test measures the proportion of samples that results in a correct decision, we want $1 - \beta$ to be large. A number of factors influence the value of $1 - \beta$, including the level of significance of the test α, the size of the sample, and the degree of the difference between the true value of the parameter and the hypothesized value.

It is important to interpret the power of the test carefully. Suppose the null hypothesis is false. Then, if we could take millions of samples of the same size n, $1 - \beta$ is interpreted as the proportion of these samples that will result in rejecting the (false) null hypothesis, that is, the chance that the test will correctly reject a false null hypothesis.

Calculating Power

In this lab, you choose a parameter D, which is a measure of the difference in the true value of the population parameter and the hypothesized value. The precise definition of D is dependent on the sampling distribution of the test statistic. Here, we give these definitions of D in terms of $\delta = D/100$ (which is simply D expressed in decimal form).

1 For the Z and one-sample t tests, the true mean of the population being sampled from is

$$\mu = \mu_0 + \delta\sigma,$$

where for normal populations, $\mu_0 = 0, \sigma = 1$; for uniform populations, $\mu = 1/2$ and $\sigma = \sqrt{1/12}$; and for exponential distributions, $\mu = \sigma = 1$.

2 For the chi-square test, the true standard deviation of the population being sampled from is

$$\sigma^2 = (1 + \delta)\sigma_0^2,$$

where $\sigma_0^2 = 1$, $1/12$, and 1 for normal, uniform, and exponential populations, respectively.

3 For the two-sample t test, the true difference in means of the populations being sampled from is

$$\mu_1 - \mu_2 = \delta\sigma,$$

where, again, $\sigma = 1, \sqrt{1/12}$, and 1 for normal, uniform, and exponential populations, respectively.

4 For the F test, the true ratio of standard deviations of the populations being sampled from is

$$\frac{\sigma_1^2}{\sigma_2^2} = 1 + \delta,$$

where $\sigma_2^2 = 1$, $1/12$, and 1 for normal, uniform, and exponential populations, respectively.

Notice in all the definitions that if $D = 0$, the null hypothesis is true, and you should get the same results from this lab as in the *Level of Significance of a Test* Lab.

The Normal Theory Power

We have used the Z, t, χ^2, and F curves to describe the sampling distributions of the statistics we are studying when the null hypothesis is true and we are sampling from a normal population. We can also find other curves that are appropriate when the null hypothesis is false, and from these curves, we can find the true power of each test for a specified value of D. We will refer throughout the guided tour to such a true value of power as the *normal theory power* of the test.

Sample Size Determination

The most important use of the idea of power is in determining how large a sample is needed for a statistical study. What we want is a sample size such that the level of significance of our test of hypothesis is a specified value and the power of the test for a specified value of D is a certain value. At the end of the Z statistic part of the guided tour and in the chapter exercises, we give examples of how to determine sample sizes.

16.2 Objectives

1. Learn the definition and proper interpretation of the power of a test.
2. Understand the effect of increasing the sample size on the power of the test.
3. Understand the effect of changing the level of significance of a test on the power of the test.
4. Understand how the size of the difference between the true value of the population parameter and the hypothesized value affects the power of the test.
5. For the five test statistics considered in the lab, know the effect of the violation of assumptions on the power of the test.

16.3 Description

To begin the lab, choose Power of a Test from Labs in the StataQuest main menu. The dialog box shown in Figure 16.1 will open, containing a brief description of the lab and six boxes requiring information.

1. n_1: The number supplied in the box to the immediate right of n_1 indicates the size of the samples to be chosen from the population if Statistic is any of Z, One-sample-t, or Chi-square. If Statistic is Two-sample-t or F, then n_1 is the size of the samples drawn from the first population. The value of n_1 should be an integer and must be between 2 and 100. The initial value of n_1 is 10.

2. n_2: The number supplied in the box to the immediate right of n_2 indicates the size of the samples to be chosen from the second population if Statistic is Two-sample-t or F. If Statistic is Z, One-sample-t,

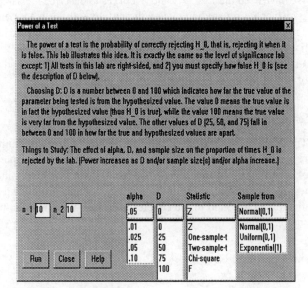

FIGURE 16.1
Dialog box for the Power of a Test Lab.

or Chi-square, the value of n_2 is ignored. The value of n_2 should be an integer and must be between 2 and 100. The initial value of n_2 is 10.

3. alpha: The number supplied in the box beneath alpha is the desired level of significance of the test. The value of alpha must be between, but not equal to, 0 and 1. If a value of alpha is desired that is not given in the choices, simply type in the desired value.

4. D: The value specified in the box directly under D is a measure of the size of the difference between the true value of the population parameter and the value specified in the null hypothesis. If D is 0, the two values are the same, so the null hypothesis is true. The larger the choice for D, the greater the difference. The values of D must be between 0 and 100. The initial value of D is 0.

5. Statistic: The choice specified for Statistic corresponds to the type of test to be performed. The initial Statistic chosen for study is the Z test statistic.

6. Sample from: The choice specified in the box beneath Sample from indicates the distribution of the population from which the samples are chosen. There are three distributions that may be selected: Normal(0,1), Uniform(0,1), or Exponential(1). All of these distributions were discussed in the *How Are Populations Distributed?* Lab. The initial distribution is Normal(0,1). The choices for Sample from are restricted to those shown.

In the bottom left-hand corner of the dialog box are three buttons.

1. Run: Clicking on this button begins the lab. A graphics window will open briefly. This window contains a status bar that fills with yellow as 500 samples of size n are drawn from the population specified in Sample from, and the value of the statistic is calculated for each sample. After the samples are selected and computations are completed, a graph similar to the one shown in Figure 16.2 will appear. This graph contains a yellow histogram of the 500 values of the test statistic with the theoretical sampling distribution superimposed in red. The

FIGURE 16.2
Theoretical distribution and histogram for a right-sided level $\alpha = 0.05$ Z test based on 500 samples of size $n = 10$ from an N(0,1) population for a true null hypothesis ($D = 0$).

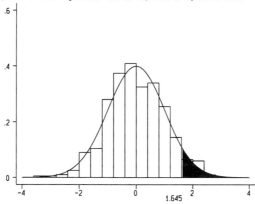

region shaded in red is the rejection region of the theoretical sampling distribution. At the top of the graph is a heading that contains the statistic for which the lab was run, the sample size, the population distribution from which samples were drawn, the value of D, the value of α, and the proportion of times that the null hypothesis was rejected for the 500 test statistics.

2 `Close`: Clicking on this button closes the lab and returns you to the StatConcepts menus.

3 `Help`: Clicking on this button opens a help window containing information about the lab.

16.4 Guided Tour of the Lab

The tours of the *Sampling Distributions* and *Level of Significance of a Test* Labs presented preliminaries for the coming attractions in the *Power of a Test* Lab. To genuinely appreciate all of the sites in this tour, which are intricately interwoven with sampling distributions and levels of significance, if you haven't already done so, take some time to study these concepts. Then, when you are ready, return and enjoy the elegant relationships contained within this lab.

16.4.1 The Z Test Statistic

Begin the lab by selecting `Power of a Test` from the `Labs` menu. The first statistic we are going to study is the Z test statistic. The Z statistic is commonly used for purposes of statistical inference on the population mean μ. If the parent population has a distribution belonging to the normal family or the sample size is large, then the sampling distribution of Z is a Normal(0,1) distribution if the null hypothesis is true. For this section of the lab, the null hypothesis will be $H_0: \mu = 0$. Thus, we are testing to see if the population is centered at zero. The logic behind hypothesis testing is to use the information contained in the data in the form of a test statistic and to assess the evidence against the null hypothesis. For the Z test, the data test H_0 by measuring the standardized distance of the sample mean $\bar{X}$ from the hypothesized value of the mean $\mu_0 = 0$, giving rise to the test statistic

$$Z = \frac{\bar{X} - \mu_0}{\sigma/\sqrt{n}} = \frac{\bar{X} - 0}{\sigma/\sqrt{n}}.$$

Notice that once a sample size has been established, the only source of variation in Z is the value of $\bar{X}$. If the population's true mean is 0, then the value of $\bar{X}$ should be close to 0, and hence the numerator of Z will also be close to zero $100(1 - \alpha)\%$ of the time. Therefore, the sampling distribution Z will be centered at zero. On the other hand, if the population's true mean is not 0, this will be reflected in the value of $\bar{X}$. The difference between the sample mean and the hypothesized value (0) will be large, resulting in the sampling distribution of the values of Z not being centered at zero.

For the first run of the lab, let `n_1 = 10`, `alpha = 0.05`, `D = 0`, `Statistic` be Z, and `Sample from` be `Normal(0,1)`. A yellow status bar will appear that fills while the lab generates 500 samples of size n_1 = 10 from the Normal(0,1) distribution and calculates the value of Z for each sample. After the calculations are complete, a graph similar to the one shown in Figure 16.2 will appear. Your graph will differ slightly because the lab generates a different set of 500 samples each time it is run. On this graph, the yellow histogram is a plot of the 500 values of Z. The red curve is the distribution of Z if the assumptions are satisfied and if the null hypothesis is true. The region shaded in red is the rejection region for a Right-Sided test with alpha = 0.05. The value 1.645 is the critical value, so that the area to the right of 1.645 is equal to alpha; $Z_{0.05} = 1.645$. (For more information on critical values, see the *Critical Values* Lab.) The caption of this graph contains information about the conditions under which the lab was run and also gives the proportion of samples that resulted in a value of Z that led to rejecting the null hypothesis (that is, values greater than 1.645).

Recall that the value of D represents the size of the difference between the actual value of the population parameter and the value of the parameter under the null hypothesis. Because D is 0, this difference is zero, and so the null hypothesis is true; that is, the actual value of the mean of the population is $\mu = 0$. Because the null hypothesis is true, if we reject it, we are making a Type I error. The value of alpha = 0.05 is the level of significance of the test. From the caption of the graph, we see that the proportion of samples that resulted in rejecting the null hypothesis is 0.052. Thus, only 5.2% of the 500 samples had "enough evidence" to (incorrectly) conclude that the null hypothesis was false. The distribution of the parent population (specified in Sample from) is a member of the normal family, so the sampling distribution of Z will be Normal(0,1). The histogram matching the theoretical curve is evidence of this. In particular, note that both the histogram and the theoretical curve are centered at zero.

Leave all values the same as for the first run of the lab, but change D to 50. By doing so, you are increasing the mean μ of the population distribution to 0.50 (so that the null hypothesis is false). Run the lab. The resulting figure will look somewhat like what is shown in Figure 16.3. The red curve is the Normal(0,1) distribution, the

FIGURE 16.3
Theoretical distribution and histogram for a right-sided level $\alpha = 0.05$ Z test based on 500 samples of size $n = 10$ from a Normal(0,1) population for a false null hypothesis $(D = 50)$.

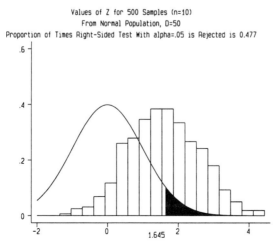

sampling distribution of the Z if the null hypothesis is true. The yellow histogram is the distribution of the 500 values of Z from a normal distribution with a mean that is larger than 0. Note that its center is not at zero. Every value of Z that is larger than 1.645 (beginning at the left-most side of the red-shaded rejection region) will result in correctly rejecting the (now false) null hypothesis. The proportion of the 500 Z's that resulted in rejecting H_0 is 0.477. This means we are making a correct decision for only 47.7% of these 500 samples. In fact, it can be shown that the actual power of the test under these circumstances is exactly equal to 0.4745. Therefore, if we could take millions of samples of size $n = 10$ from a Normal(0.5,1) distribution, the proportion of samples resulting in a value of $Z > 1.645$ would be 0.4745.

If we are rejecting a false null hypothesis only a little more than 47% of the time, a natural question would be, "How can this percentage be increased?" Let the sample size n_1 = 50 and run the lab again. The result is shown in Figure 16.4. This increase in sample size drastically improved the power of the test! The proportion of samples resulting in a test statistic that rejects H_0 is now 0.968. The theoretical power of the test in this case is 0.9707. Thus, if we determine for a given sample size that the test we will be performing has a low power, one way to increase the power would be to increase the sample size n.

What if resources are unavailable to increase the sample size? Is there another way to increase the power of the test? Change n_1 back to 10 and increase alpha to 0.10. Run the lab. It can be shown that the normal theory power for this case is 0.6176. In Figure 16.5, we see that the proportion of the 500 samples that actually rejected H_0 is 0.659. Thus, we see that there is another way to increase the power of the test. Unfortunately, it also increases the likelihood of a Type I error.

Finally, let's see what happens to the power of the test as the distance between the null value of μ and the actual value of μ increases. Change alpha back to 0.05 and let D = 100, so that the true value of the population mean is 1. Run the lab. In Figure 16.6, we see that the proportion of values from the 500 samples that were larger than 1.645 is 0.940. The normal theory power of the test is 0.9354. What we learn from this is that the larger the difference between μ_0 and μ, the more certain we can be that we will detect this difference.

FIGURE 16.4
Theoretical distribution and histogram for a right-sided level $\alpha = 0.05$ Z test based on 500 samples of size $n = 50$ from a Normal(0,1) population for a false null hypothesis ($D = 50$).

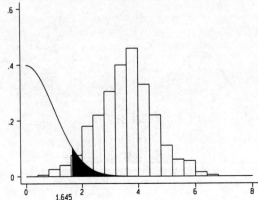

FIGURE 16.5
Theoretical distribution and histogram for a right-sided level $\alpha = 0.10$ Z test based on 500 samples of size $n = 10$ from a Normal(0,1) population for a false null hypothesis ($D = 50$).

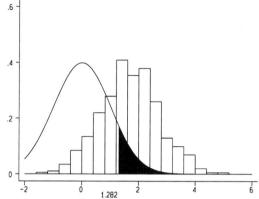

FIGURE 16.6
Theoretical distribution and histogram for a right-sided level $\alpha = 0.05$ Z test based on 500 samples of size $n = 10$ from a Normal(0,1) population for a false null hypothesis ($D = 100$).

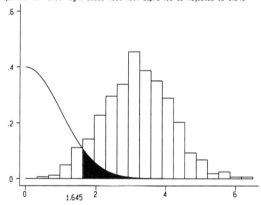

You should now have a good grasp of how the power of a Z test behaves when the normality assumption is met. But how is power affected when the population distribution is not normal? Before we answer this question, run the lab for `n_1 = 10`, `alpha = 0.05`, `D = 0`, and `Sample from of Uniform(0,1)`. For a Uniform(0,1) parent population, the null hypothesis is H_0: $\mu = 0.50$. Because the variance of a Uniform(0,1) distribution is 1/12, the test statistic becomes

$$Z = \frac{\bar{X} - \mu_0}{\sigma/\sqrt{n}} = \frac{\bar{X} - 0.50}{(1/12)/\sqrt{n}}.$$

Recall from the *Sampling Distributions* and *Level of Significance of a Test* Labs that when the null hypothesis is true, the agreement between the theoretical Normal(0,1) curve and the histogram of the 500 values of the test statistic is very good, even though the population has a uniform distribution. For reference, we show you this again in Figure 16.7.

FIGURE 16.7
Theoretical distribution and histogram for a right-sided level $\alpha = 0.05$ Z test based on 500 samples of size $n = 10$ from a Uniform(0,1) population for a true null hypothesis ($D = 0$).

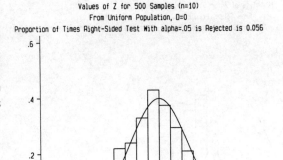

Now change D to 50. The result should look similar to what is shown in Figure 16.8. Using the expressions in Subsection 16.1.1, we find that for this value of D, the actual value of the population mean is $\mu = \mu_0 + \delta\sigma = 0.5 + (0.5)(\sqrt{1/12}) = 0.6443$. If the normal approximation to the sampling distribution of Z is good, then the power of the test (the normal theory power) should be very close to 0.4745. In Figure 16.8, the observed proportion of samples that resulted in a value of $Z > 1.645$ is 0.485. Thus, again, we see that if the population has a uniform distribution, even a small sample size is large enough to confidently use the normal distribution for purposes of statistical inference. If you ran the lab for similar cases as we studied for the normally distributed parent population, you would witness the same types of behavior in terms of both the power of the test and the closeness of the normal distribution to the sampling distribution of Z.

Can the same thing be said about an exponentially distributed parent? Change D back to 0, and `Sample from` to `Exponential(1)`. Run the lab. For this case,

FIGURE 16.8
Theoretical distribution and histogram for a right-sided level $\alpha = 0.05$ Z test based on 500 samples of size $n = 10$ from a Uniform(0,1) population for a false null hypothesis ($D = 50$).

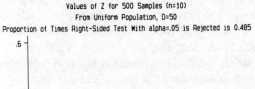
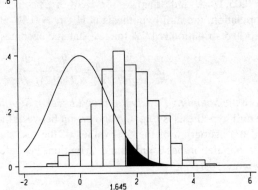

the null hypothesis is $H_0: \mu = 1$, and the test statistic is

$$Z = \frac{\bar{X} - \mu_0}{\sigma/\sqrt{n}} = \frac{\bar{X} - 1}{1/\sqrt{n}}.$$

The graph in Figure 16.9 should remind you that for an exponentially distributed parent population, the sample size needs to be larger for the sampling distribution of Z to be well approximated by a standard normal distribution. The actual proportion of Z values that resulted in a Type I error was larger than the desired 0.05.

How does this poor approximation affect the power of the test? Change D to 50 and run the lab again to find out. For this value of D, the true value of the population mean is 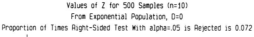 In Figure 16.10, compare the proportion of the 500 samples that rejected H_0 (0.429) to 0.4745. Not only has an exponentially distributed parent combined with a small sample size hurt our Type I error rate, but the Type II error rate is also compromised! Figure 16.11 shows the result for the same values but for a sample of size n_1 = 50. The power of the test under these circumstances should be very close to 0.9707 if the normal approximation has improved with the increase in sample size. The proportion of 500 samples that rejected H_0 in our result was 0.986. Therefore, we see if the sample size is large enough, the normal approximation is reliable for an exponentially distributed parent.

FIGURE 16.9
Theoretical distribution and histogram for a right-sided level $\alpha = 0.05$ Z test based on 500 samples of size $n = 10$ from an Exponential(1) population for a true null hypothesis ($D = 0$).

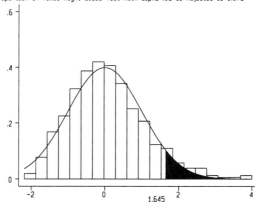

16.4.2 Choosing Sample Sizes

One important use of power is in determining what sample size should be used in a statistical study. To illustrate this, we'll run the lab to determine what value of n will result in a power of 0.90 for $\alpha = 0.05$ and $D = 50$ when we are sampling from a normal population. Thus, we want to be 90% sure we can conclude that the population mean is not 0 when it is really 0.5 and also be 95% sure that we can conclude that the mean is 0 when it actually is 0.

If we run the lab several times for n_1 = 10, we find that between 45 and 50% of the time, we (correctly) reject the null hypothesis. Thus, we need to increase n_1. If we use 20 and run the lab several times, we reject between 70 and 75% of the time.

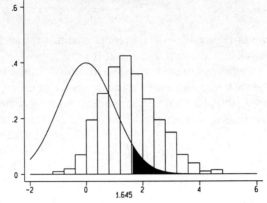

FIGURE 16.10
Theoretical distribution and histogram for a right-sided level $\alpha = 0.05$ Z test based on 500 samples of size $n = 10$ from an Exponential(1) population for a false null hypothesis ($D = 50$).

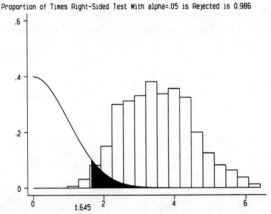

FIGURE 16.11
Theoretical distribution and histogram for a right-sided level $\alpha = 0.05$ Z test based on 500 samples of size $n = 50$ from an Exponential(1) population for a false null hypothesis ($D = 50$).

For n_1 = 30, the rejection rates are between 83 and 87%. Finally, n_1 = 35 gives rates between 88 and 92%. Therefore, we conclude that a sample size around 35 will give our test the level of significance and power we want.

If we repeat this whole process for $D = 100$, we can expect to need a smaller sample size because now we are trying to tell the difference between means of 0.0 and 1.0, rather than 0.0 and 0.50 as in the $D = 50$ case. In fact, we find that we need only a sample of size 8 for the power to be in the neighborhood of 90%.

This process of sample size determination can be done at each stop of the guided tour, but we leave it for the exercises.

16.4.3 The One-sample t Test Statistic

The one-sample t statistic is also used for purposes of statistical inference on the population mean when the value of the population variance σ^2 is unknown and is estimated by s^2. For a null hypothesis $H_0: \mu = \mu_0$, this results in the test statistic

$$t = \frac{\bar{X} - \mu_0}{s/\sqrt{n}}.$$

Having to estimate σ^2 introduces more variability in the possible values of the test statistic because two samples with the same value of $\bar{X}$ can have different values of s^2. We saw in the *Sampling Distributions* Lab that if the parent population has a distribution belonging to the normal family, the shorter peak and thicker tails of the t_ν distribution explain this added variation.

If you have not already done so, change the value of Statistic to One-sample-t. We begin our study of the power of the test for the one-sample t test by running the lab for n_1 = 10, alpha = 0.05, D = 0 (so that the null hypothesis is true), and Sample from of Normal(0,1). Because the parent population is normally distributed, the normality assumption is met. For a normal parent, in this lab, the null hypothesis is $H_0: \mu = 0$. Run the lab for the given values. The result you get should look something like Figure 16.12. Notice the agreement between the centers (and shapes) of the two distributions displayed on the graph. Because the null hypothesis is true, rejecting it would be a Type I error. We have specified that we would like the chance of that occuring to be 0.05, and in the graph, the proportion of samples resulting in a Type I error is 0.05.

Now change the value of D to 50 and run the lab. The result is shown in Figure 16.13. Here, we stress that the null hypothesis remains $H_0: \mu = 0$. For this value of D, the value of the mean of the population is $\mu = \mu_0 + \delta\sigma = 0 + (0.50)(1) = 0.50$. You may recall from the Z, t, χ^2, and F Lab that all t_ν distributions have mean 0. Now, if the value of μ is not zero, then the sample mean will reflect this. Therefore, the numerator of the one-sample t test statistic will be "far" from zero. In this case, the sampling distribution of the t statistic will not be centered at zero. The shift to the right of the histogram of the 500 one-sample t statistics is an example of this. Because the location of the distribution is affected by a change in mean, the name of this new distribution is, appropriately, the *noncentral t distribution* with degrees of

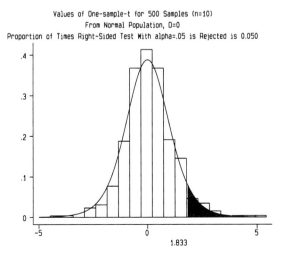

FIGURE 16.12 Theoretical distribution and histogram for a right-sided level $\alpha = 0.05$ one-sample t test based on 500 samples of size $n = 10$ from a Normal(0,1) population for a true null hypothesis ($D = 0$).

FIGURE 16.13 Theoretical distribution and histogram for a right-sided one-sample level $\alpha = 0.05$ t test based on 500 samples of size $n = 10$ from a Normal(0,1) population for a false null hypothesis $(D = 50)$.

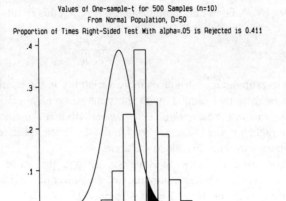

freedom $n - 1$. The shape and spread of this distribution is also affected. Expressions for the mean and variance of the noncentral t distribution are a somewhat complicated function of the degrees of freedom and a parameter called the *noncentrality parameter*, expressed as $\delta = D/100$. In Figure 16.13, the histogram is slightly shifted to the right and skewed to the right. The observed power of the test in this graph is 0.411. For a sample of size $n\_1 = 10$ and a true mean $\mu = 0.50$, the percentage of times the 500 samples rejected H_0 is only 41.1%.

In the section on the Z statistic, we saw how increasing the sample size will also increase the power of the test. To examine what will happen in the case of the one-sample t statistic, increase the value of n_1 to 50. Figure 16.14 shows that increasing the sample size did indeed increase the power. The proportion of samples resulting in a Type II error for $D = 50$ and a sample of $n\_1 = 50$ is 0.972—a huge improvement.

We also saw that the farther away the true value of the population mean was from the hypothesized value, the more certain we could be that we would find that

FIGURE 16.14 Theoretical distribution and histogram for a right-sided level $\alpha = 0.05$ one-sample t test based on 500 samples of size $n = 50$ from a Normal(0,1) population for a false null hypothesis $(D = 50)$.

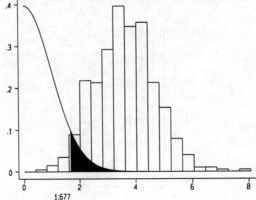

difference. To see this for the One-sample-t statistic, change n_1 back to 10 and let D = 100. The true value of the population mean is $\mu = 1$. Run the lab. The result is shown in Figure 16.15. Not only has the noncentral t distribution shifted farther to the right, it has also become more skewed than for D = 50. These two properties combined explain why the proportion of samples resulting in a value of t larger than $t_{0.05,9} = 1.833$ is 0.906! Comparing this to a value of 0.411 for D = 50, you can see that there is quite an improvement. Before we examine how the power of the test behaves when the normality assumption is not satisfied, it is worth mentioning that if you increase the value of alpha, then the power of the test will also increase. This particular property was shown for the Z statistic and remains the same for the one-sample t because in both cases, by increasing alpha, we make it easier to reject the null hypothesis.

Now, how does non-normality affect the results of statistical inference based on a t_ν distribution? Change D back to zero and let Sample from be Uniform(0,1). Run the lab. Figure 16.16 illustrates one possible result. After examining this figure,

FIGURE **16.15**
Theoretical distribution and histogram for a right-sided level $\alpha = 0.05$ one-sample t test based on 500 samples of size $n = 10$ from a Normal(0,1) population for a false null hypothesis ($D = 100$).

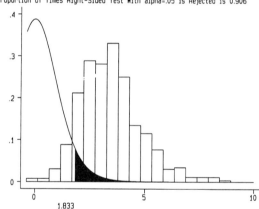

FIGURE **16.16**
Theoretical distribution and histogram for a right-sided level $\alpha = 0.05$ one-sample t test based on 500 samples of size $n = 10$ from a Uniform(0,1) population for a true null hypothesis ($D = 0$).

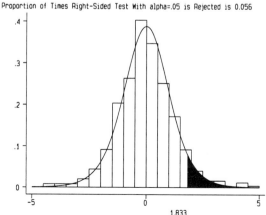

you should recall that the t_ν distribution also works well as the sampling distribution of the one-sample t statistic if the population has a uniform distribution. The specified value for the proportion of samples resulting in a Type I error was 0.05. The actual proportion of the 500 samples is 0.056.

Figure 16.17 shows the result of the lab for D = 50. Note that the proportion of 500 samples that results in rejecting the false null hypothesis is disconcertingly small. That the population distribution is not normal has affected the goodness of using the noncentral t to find the power of the test. If the noncentral t was the appropriate distribution, the power of this test for these values would be approximately 0.40. But in our set of 500 samples, we only had 0.377. Therefore, although the Type I error rate has not been compromised by the non-normality of the parent distribution, the power of the test has. If we increase the sample size to $n_{-}1 = 50$, the power of the test should be approximately 0.97. In Figure 16.18, we see that increasing the sample

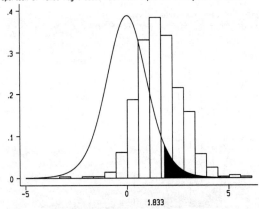

FIGURE **16.17** Theoretical distribution and histogram for a right-sided level $\alpha = 0.05$ one-sample t test based on 500 samples of size $n = 10$ from a Uniform(0,1) population for a false null hypothesis ($D = 50$).

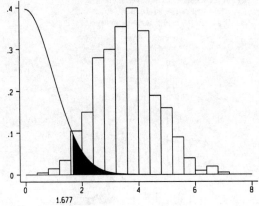

FIGURE **16.18** Theoretical distribution and histogram for a right-sided level $\alpha = 0.05$ one-sample t test based on 500 samples of size $n = 50$ from a Uniform(0,1) population for a false null hypothesis ($D = 50$).

size to 50 did indeed result in a power of approximately 0.97 and that the lack of normality in the parent population no longer poses much of a problem.

We are now going to examine the power of the test for an exponential population distribution. As before, we will run the lab first for the case in which the null hypothesis is true. In Figure 16.19, we see that the actual Type I error rate is much smaller than the specified 0.05. This is the first indication that the sampling distribution of the one-sample t statistic under an exponential parent is not well modeled by the t_9 distribution. But at the present time, that doesn't look like much of a problem, because we have reduced the Type I error rate. Figure 16.20 tells the rest of the story. The power of the test (0.343) in this figure is lower than it should be, which is around 0.40. Thus, again, we see that deviations from the normality assumption jeopardize the validity of the statements we can make about the certainty of our inference. If you run the lab for an increased sample size of 50, just as in other cases, you will find that the power of the test increases dramatically and the lack of normality ceases to be a factor.

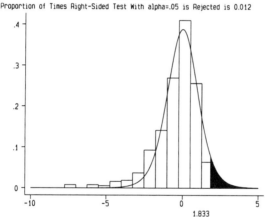

FIGURE **16.19**
Theoretical distribution and histogram for a right-sided level $\alpha = 0.05$ one-sample t test based on 500 samples of size $n = 10$ from an Exponential(1) population for a true null hypothesis ($D = 0$).

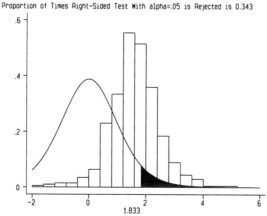

FIGURE **16.20**
Theoretical distribution and histogram for a right-sided level $\alpha = 0.05$ one-sample t test based on 500 samples of size $n = 10$ from an Exponential(1) population for a false null hypothesis ($D = 50$).

16.4.4 The Two-sample t Test Statistic

The two-sample t statistic is used for comparing the means of two populations. The two-sample t statistic has a $t_{n_1+n_2-2}$ distribution if both populations have normal distributions and are independent of each other, if the values of the population variances are equal, and if the null hypothesis is true. If all other assumptions are satisfied but the null hypothesis is false, then the distribution of the two-sample t statistic is a noncentral t distribution with degrees of freedom $n_1 + n_2 - 2$ and noncentrality parameter δ. In all cases we will consider in this lab, the null hypothesis will be that there is no difference in the means: $H_0: \mu_1 - \mu_2 = 0$. Therefore, the test statistic is given by the expression

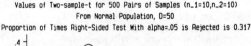

$$t = \frac{(\bar{X}_1 - \bar{X}_2) - 0}{\sqrt{s_p^2 \left(\frac{1}{n_1} + \frac{1}{n_2} \right)}}.$$

Because we spent a lot of time in the previous section examining the behavior of the t distribution for various circumstances, we will spend less time on it here. However, we will look at a few examples so you can see what to expect for the two-sample t. Because the accuracy of the Type I error can be seen for different circumstances in Figures 16.12, 16.16, and 16.19, we will not run the lab for D = 0 in this section. Before we begin, change the value of Statistic to Two-sample-t.

The first case we consider in this section is the case in which the populations being sampled from are Normal(0,1), n_1 = n_2 = 10, alpha = 0.05, and D = 50. The result is shown in Figure 16.21. For this particular value of D, the difference between the two population means is $\delta \sigma = (0.50)(1) = 0.50$. Although the assumptions are met in this case, the power of this test is distressingly low; only 31.7% of the samples resulted in rejecting the false null hypothesis. There are two ways to rectify this situation. The first alternative is to raise the probability of a Type I error.

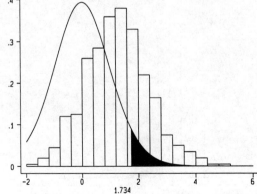

FIGURE **16.21**
Theoretical distribution and histogram for a right-sided level $\alpha = 0.05$ two-sample t test based on 500 samples of size $n_1 = n_2 = 10$ from Normal(0,1) populations for a false null hypothesis $(D = 50)$.

Let alpha = 0.10 and run the lab again. We are now willing to take a 10% chance of rejecting the null hypothesis if it happens to be true. In Figure 16.22, we see that an increase of 5% in alpha resulted in an approximate increase of 10% in the power of the test. However, we find a difference that exists for 41.7% of the samples is still quite low. The second alternative we have is to increase the sample size.

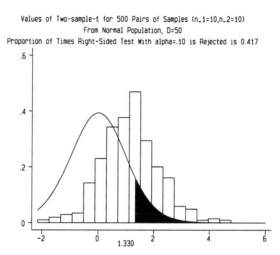

FIGURE 16.22
Theoretical distribution and histogram for a right-sided level $\alpha = 0.10$ two-sample t test based on 500 samples of size $n_1 = n_2 = 10$ from Normal(0,1) populations for a false null hypothesis ($D = 50$).

For reference, change alpha back to 0.05 and let n_1 = n_2 = 50. This is quite a large jump, from observing a total of 20 elements to 100. How will such a large increase affect the power of the test? We would expect that for such a large amount of information, the power would be quite high. The graph in Figure 16.23 indicates that it is 0.794. The same type of increase in the one-sample situation caused a much larger

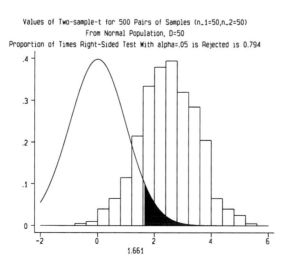

FIGURE 16.23
Theoretical distribution and histogram for a right-sided level $\alpha = 0.05$ two-sample t test based on 500 samples of size $n_1 = n_2 = 50$ from Normal(0,1) populations for a false null hypothesis ($D = 50$).

jump in the power. The two-sample t test needs a larger sample size to pay off more because it already starts off with a larger value for the rejection region. The larger the critical value, the easier it is to reject H_0, and the larger the degrees of freedom are to begin with, the smaller the change in the rejection region when they are increased. Notice that in the two-sample case, the critical value decreased only from 1.734 (for samples of size 10) to 1.661 (for samples of size 50). For the one-sample case, a similar increase changed the rejection region from 1.833 (for samples of size 10) to 1.677. The other factor influencing the slower increase is the intricacies of how the location and shape of the noncentral t distribution are affected by changes in degrees of freedom and the noncentrality parameter. Recall that the mean and the variance of the noncentral t distribution are complex functions of the degrees of freedom and the noncentrality parameter. Because of the inherently larger degrees of freedom in the two-sample t, the nature of these functions makes increasing power more difficult.

You may be asking, "Will the power of a two-sample t test ever be large?" Change D to 75 and run the lab again. Now the difference in the population means is 0.75. In Figure 16.24, we see the answer to the question is "Yes!" But it takes a larger difference for the power of the test to be large. In Figure 16.24, the proportion of 500 samples that resulted in rejecting the false null hypothesis is a reassuring 0.978.

By now, you should be wondering, "What happens to the power of the test if the populations don't have distributions that are members of the normal family?" Figures 16.25 and 16.26 seek the answer to this question for the Uniform(0,1) and Exponential(1) distributions, respectively. In both of these figures, n_1 = n_2 = 10, alpha = 0.05, and D = 50. Recall that for the Uniform(0,1) distribution, the value of the actual Type I error was quite reliable. But in Figure 16.25, we see that the power of the test is not only low (only 27.9% of the 500 samples rejected H_0), but it is even lower than the power when the assumptions are satisfied. Therefore, having a uniformly distributed parent population compromised the power of the test. In Figure 16.26, the power of the test is a little higher than when the assumptions are satisfied.

FIGURE **16.24**
Theoretical distribution and histogram for a right-sided level $\alpha = 0.05$ two-sample t test based on 500 samples of size $n_1 = n_2 = 50$ from Normal(0,1) populations for a false null hypothesis $(D = 75)$.

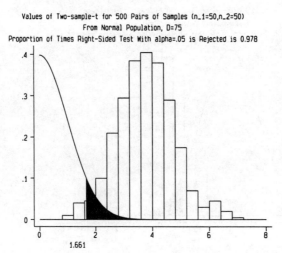

FIGURE 16.25
Theoretical distribution and histogram for a right-sided level $\alpha = 0.05$ two-sample t test based on 500 samples of size $n_1 = n_2 = 10$ from Uniform(0,1) populations for a false null hypothesis ($D = 50$).

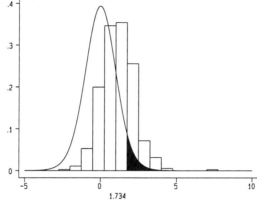

FIGURE 16.26
Theoretical distribution and histogram for a right-sided level $\alpha = 0.05$ two-sample t test based on 500 samples of size $n_1 = n_2 = 10$ from Exponential(1) populations for a false null hypothesis ($D = 50$).

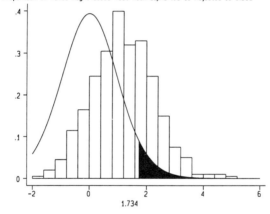

You should now run the lab for both parent distributions, increasing the sample sizes to 50. You will see that, just as before, this increases the power to results comparable to the power for normally distributed parent populations with samples of size 50. Finally, increasing D to 75 will also markedly increase the power for large sample sizes, just as it did in the case of the normally distributed parent.

16.4.5 The χ^2 Test Statistic

When testing the null hypothesis $H_0: \sigma^2 = \sigma_0^2$, if H_0 is true, the χ^2 statistic has a χ^2_{n-1} distribution if the population is normally distributed. The χ^2 test statistic is given by

$$\chi^2 = (n-1)\frac{s^2}{\sigma_0^2}.$$

In the following discussion, for a parent population with a normal distribution, $\sigma_0^2 = 1$; for a uniform distribution, $\sigma_0^2 = 1/12$; and for an exponential distribution, $\sigma_0^2 = 1$. In all cases, the effect of the value of D on the value of the variance of the population from which the sample is being selected can be described by

$$\sigma^2 = [1 + (D/100)]\sigma_0^2.$$

A positive value for D will result in an increase in the variance of the population. This, in turn, will cause the variance of the values of χ^2 to increase, so that if the null hypothesis is false, the distribution of the χ^2 statistics will be flatter and more spread out than the χ_{n-1}^2 distribution. Notice that if $D = 0$, the null hypothesis is true.

If the normality assumption is satisfied, we know from previous labs that the sampling distribution of the χ^2 test statistic is exactly a χ_{n-1}^2 distribution. We will now examine what happens to the sampling distribution if H_0 is false. If you have not already done so, change the value of Statistic to Chi-square. Let n_1 = 10, alpha = 0.05, D = 50, and Sample from be Normal(0,1). For this parent distribution, the null hypothesis in the lab is H_0: $\sigma^2 = 1$. For D = 50, the value of the population variance is $\sigma^2 = (1 + 0.5)(1) = 1.50$. Further, it can be shown that the normal theory power of the test under these conditions is 0.2570. Run the lab. The resulting figure should resemble Figure 16.27. Note that the histogram of the 500 χ^2 statistics is shorter and has a longer right tail than the χ_9^2 distribution. This is because the variance of the population is larger than the hypothesized value. Note also that the proportion (0.253) of the 500 samples is very close to the actual power of the test (0.2570). However, it is disturbingly low.

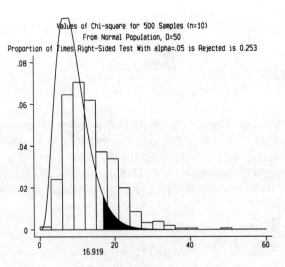

FIGURE **16.27**
Theoretical distribution and histogram for a right-sided level $\alpha = 0.05$ χ^2 test based on 500 samples of size $n = 10$ from a Normal(0,1) population for a false null hypothesis $(D = 50)$.

In light of this second realization, we naturally want to increase the power of this test. Increase the sample size to n_1 = 50 and run the lab again. The result is shown in Figure 16.28. For this increased sample size, the normal theory power

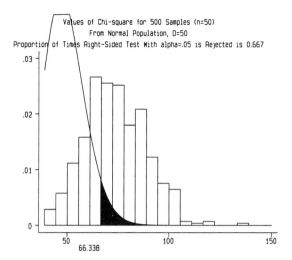

FIGURE **16.28**
Theoretical distribution and histogram for a right-sided level $\alpha = 0.05$ χ^2 test based on 500 samples of size $n = 50$ from a Normal(0,1) population for a false null hypothesis ($D = 50$).

of the test has increased to 0.6668. The figure confirms this. The proportion of the 500 samples resulting in a correct decision is 0.667. This is still lower than most reasonable analysts would insist upon. Much larger sample sizes are required for a large power for inference on the population variance. Even for a sample size as large as $n = 100$, the power of the test when $D = 50$ is only 0.8897.

Will increasing alpha help much? Run the lab for alpha = 0.10. For these values the power is 0.7727, an increase of 0.1059. In Figure 16.29, we see that the proportion of 500 samples resulting in rejecting H_0 is 0.776. If you pull all these facts together, you see that for the power of the χ^2 test to be reasonably high, both a larger sample size and a larger value of α are required.

Before we examine how violation of the normality assumption affects the power of the χ^2 test, let's see how a larger value of D affects the power. Change alpha

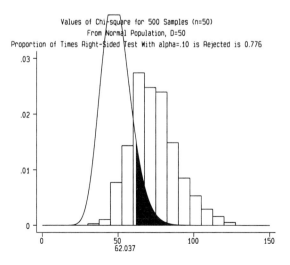

FIGURE **16.29**
Theoretical distribution and histogram for a right-sided level $\alpha = 0.10$ χ^2 test based on 500 samples of size $n = 50$ from a Normal(0,1) population for a false null hypothesis ($D = 50$).

back to 0.05. By increasing the value of D to 100, we are increasing the value of the population variance to $\sigma^2 = (1 + 1)(1) = 2$, which, in turn, theoretically increases the power of the test to 0.9594. Figure 16.30 exhibits this. The true population variance is large enough that the spread of the histogram that represents the actual sampling distribution of the χ^2 statistics is very wide. Further, the proportion of these statistics correctly rejecting the null hypothesis is 0.958. Thus, although small differences between a population variance and a hypothesized value may be difficult to find consistently, larger differences are reliably detected for a large enough sample size.

Recall from the *Sampling Distributions* and *Level of Significance of a Test* Labs that if the parent population of a χ^2 test statistic is the Uniform(0,1) distribution, the actual level of the test is much lower than the specified value. If this is so, then how is the power of the test affected if the population sampled from has a Uniform(0,1) distribution? To find out, change n_1 back to 10 and D back to 50, let Sample from be Uniform(0,1), and run the lab. For these particular values, if the non-normality has no effect, then the power should be approximately 0.2570. But in Figure 16.31 we see instead that the already low power of the test is even lower. Only 20.4% of the 500 samples resulted in rejecting the null hypothesis. However, it should be noted that as the sample size increases, this discrepancy is somewhat quickly overcome. In fact, the power under a uniform distribution actually beats the power for normal populations for a large enough n.

You also might remember that when a parent population has an exponential distribution, the observed value of the Type I error is higher than desired. This will have a positive effect on the power of the test. Change Sample from to Exponential(1) and run the lab. The result is shown in Figure 16.32. Notice that the proportion of samples resulting in a value of the χ^2 test statistic that is greater than 16.919 (so that H_0 is rejected) is 0.267, just a little larger than the original 0.2570 (under normality of the parent). You can run the lab for larger sample sizes to see how quickly the differences become negligible. But you will find that as the sample

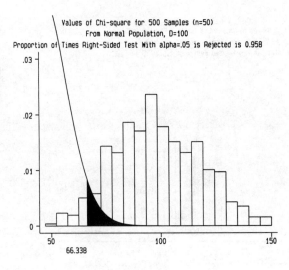

FIGURE 16.30
Theoretical distribution and histogram for a right-sided level $\alpha = 0.05$ χ^2 test based on 500 samples of size $n = 50$ from a Normal(0,1) population for a false null hypothesis ($D = 100$).

FIGURE **16.31**
Theoretical distribution and histogram for a right-sided level $\alpha = 0.05$ χ^2 test based on 500 samples of size $n = 10$ from a Uniform(0,1) population for a false null hypothesis ($D = 50$).

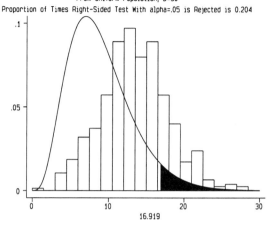

FIGURE **16.32**
Theoretical distribution and histogram for a right-sided level $\alpha = 0.05$ χ^2 test based on 500 samples of size $n = 10$ from an Exponential(1) population for a false null hypothesis ($D = 50$).

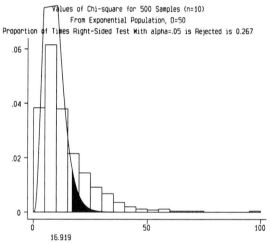

size gets larger and larger, the power actually suffers for an exponential parent when compared to the power for a normal parent.

16.4.6 The F Test Statistic

The final test statistic we will study is the F statistic, which is used to compare two population variances. The F statistic is best known as the test statistic in inference procedures in analysis of variance and regression. In this section, the null hypothesis is $H_0: \sigma_1^2/\sigma_2^2 = 1$. If this null hypothesis is true, and if the populations are independent and have distributions belonging to the normal family, then the F test statistic, given by

$$F = \frac{s_1^2}{s_2^2},$$

has an F_{n_1-1,n_2-1} distribution. The effect of the value of D on the ratio of the population variances is expressed by

$$\frac{\sigma_1^2}{\sigma_2^2} = 1 + D/100.$$

Note that if $D = 0$, the ratio is 1, and the null hypothesis is true. Otherwise, the ratio of the variances is larger than 1, and so the sampling distribution of the F statistics will have a longer right tail and be shorter than the null F_{n_1-1,n_2-1} distribution.

In the *Sampling Distributions* and *Level of Significance of a Test* Labs, we saw similarities in the χ^2 and F distributions. We will now explore those similarities for the power of the F test. To see how the power of an F test behaves for moderate values of D and relatively small samples sizes, let n_1 = n_2 = 10 and D = 50. If you have not already done so, let Statistic be F, Sample from be Normal(0,1), and, as usual, alpha = 0.05. For these particular values, it can be shown that the normal theory power of the test is only 0.1392. The graph you get should look a lot like the one shown in Figure 16.33. Notice that the distribution of the 500 F statistics is much flatter and longer in the right tail than the $F_{9,9}$ distribution. The proportion of the 500 F statistics that were larger than 3.179 is 0.132. Even though it is close to the theoretical power, this is really low. Only 13.2% of the values of F correctly rejected that the two population variances were equal.

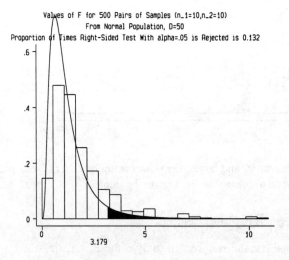

FIGURE **16.33**
Theoretical distribution and histogram for a right-sided level $\alpha = 0.05$ F test based on 500 samples of sizes $n_1 = n_2 = 10$ from Normal(0,1) populations for a false null hypothesis ($D = 50$).

Time to increase the sample sizes! Let n_1 = n_2 = 50 and run the lab again. For these sample sizes, the theoretical power of the F test is only 40.52%—still not very high, but better. Will the F statistics behave accordingly? Figure 16.34 says they do. Of the 500 F statistics, the proportion that rejected H_0 is 0.399. The graph shown in Figure 16.35 is for the same sample sizes, but alpha has been increased to 0.10. For these values, the normal theory power of the test is 0.5504. As the graph shows, the proportion of the 500 test statistics that rejected H_0 is 0.563.

FIGURE **16.34**
Theoretical distribution and histogram for a right-sided level $\alpha = 0.05$ F test based on 500 samples of sizes $n_1 = n_2 = 50$ from Normal(0,1) populations for a false null hypothesis $(D = 50)$.

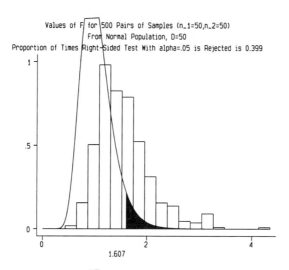

FIGURE **16.35**
Theoretical distribution and histogram for a right-sided level $\alpha = 0.10$ F test based on 500 samples of sizes $n_1 = n_2 = 50$ from Normal(0,1) populations for a false null hypothesis $(D = 50)$.

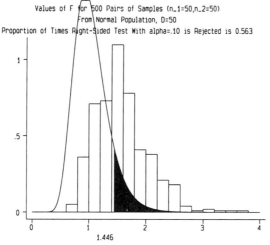

With barely over 50% of the F statistics rejecting H_0 for samples of size 50 and a significance level of 0.10, it will be interesting to see whether a larger value of D improves the power significantly. If the value of D is increased to 100, the normal theory power of the F test is 0.8699. Figure 16.36 illustrates this nicely, because the proportion of samples resulting in correctly rejecting the null hypothesis for these 500 samples is 0.874. Notice how the $F_{49,49}$ distribution is much taller than the sampling distribution of the F statistics. That is because for $D = 100$, σ_1^2 is twice as large as σ_2^2.

For a uniformly distributed parent population, we saw in the *Level of Significance of a Test* Lab that the actual value of the proportion of F statistics that results in rejecting a true H_0 is smaller than the specified level. Recall that although this looks encouraging, it might affect the power of the test. To see just how the power of the test is affected, let `Sample from` be `Uniform(0,1)`, `D = 50`, `alpha = 0.05`, and

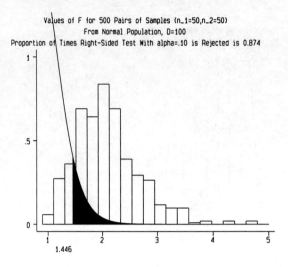

FIGURE **16.36**
Theoretical distribution and histogram for a right-sided level $\alpha = 0.10$ F test based on 500 samples of sizes $n_1 = n_2 = 50$ from Normal(0,1) populations for a false null hypothesis ($D = 100$).

n_1 = n_2 = 10. Run the lab. If the power of the test is unaffected by the non-normality of the parent population, then it will be around 0.1392. In fact, in Figure 16.37, we see that the proportion of 500 samples that reject the null hypothesis is around 0.048. This is depressingly smaller than 0.1392. However, if you continue to increase the sample sizes, you will see that the power under the uniformly distributed parent will eventually outperform the power of the normally distributed parent.

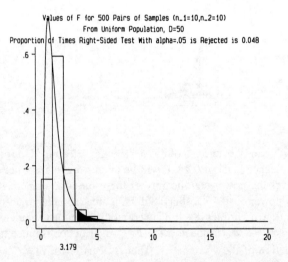

FIGURE **16.37**
Theoretical distribution and histogram for a right-sided level $\alpha = 0.05$ F test based on 500 samples of sizes $n_1 = n_2 = 10$ from Uniform(0,1) populations for a false null hypothesis ($D = 50$).

Our final graph illustrates what happens to the power of the test when the parent distribution is Exponential(1). If the F statistics are behaving as they should, the power of the F test for D = 50, n_1 = n_2 = 10, and alpha = 0.05 will be approximately 0.1392. In fact, the graph in Figure 16.38 contains 500 samples, with 22.8% rejecting the null hypothesis! This is larger than the power calculated under the normally distributed parent. Unfortunately, as the sample size increases, the power of the exponential parent will be surpassed by the power of the normal parent.

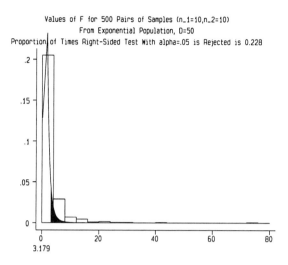

FIGURE 16.38
Theoretical distribution and histogram for a right-sided level $\alpha = 0.05$ F test based on 500 samples of sizes $n_1 = n_2 = 50$ from Exponential(1) populations for a false null hypothesis $(D = 50)$.

16.5 Summary

We observed many properties of the power of the test on this tour:

1. A Type II error is failing to reject a false null hypothesis. Suppose the null hypothesis is false. Then, if we draw all possible samples of size n from the population, the value of β is the proportion of those samples that result in failing to reject the null hypothesis.

2. The power of the test is equal to $1 - \beta$. Suppose the null hypothesis is false. Then, if we draw all possible samples of size n from the population, the power of the test is the proportion of those samples that would result in rejecting the null hypothesis.

3. If all other testing conditions remain the same, then an increase in the sample size will result in an increase in the power of the test. We use power to determine sample sizes for statistical studies.

4. If all other testing conditions remain the same, then an increase in α will result in an increase in the power of the test.

5. If all other testing conditions remain the same, then an increase in D will result in an increase in the power of the test.

6. For a Z test, a uniformly distributed parent requires only a small sample size for the normal distribution to be an excellent approximation for the sampling distribution of Z in terms of both the Type I and Type II error rates. However, a larger sample size must be used for an exponentially distributed parent.

7. For a one-sample or two-sample t test, when the null hypothesis is false, the sampling distribution of the t test statistic is a noncentral t distribution. The parameters of this distribution are the degrees of freedom ($n - 1$ for one-sample, $n_1 + n_2 - 2$ for two-sample) and the noncentrality parameter, $\delta\sigma$. The expressions for the mean and variance of a noncentral t distribution are complicated functions of the degrees of freedom and noncentrality parameter.

284 Chapter 16 Power of a Test

8 For a one-sample or two-sample t test, a uniformly distributed parent results in an accurate value for the Type I error rate but a low value for a Type II error rate. A moderate increase in sample size will alleviate the discrepancy. The same may be said for an exponential parent.

9 For a two-sample t test, a more marked increase in the sample sizes is required for the same increase in power for a one-sample t test.

10 For a χ^2 test with moderate values of D (like 25, 50, or 75), larger sample sizes and a larger value of α are required for a high power.

11 In a χ^2 test, if the parent population has a distribution belonging to the uniform family, the actual level of the test is much smaller than what is specified. The effect of this is that the power of the test is lower. The situation is reversed if the parent population is exponentially distributed; that is, the actual level of significance is higher, but the power is also higher.

12 For an F test with moderate values of D (like 25, 50, or 75), larger sample sizes and a larger value of α are required for a high power.

13 In an F test, if the parent population has a distribution belonging to the uniform family, the actual level of the test is much smaller than what is specified. The effect of this is that the power of the test is lower. The situation is reversed if the parent population is exponentially distributed; that is, the actual level of significance is higher, but the power is also higher.

16.6 Lab Exercises

1 What is a Type II error?
2 What is the difference between a Type II error and β?
3 What is the power of the test?

16.6.1 The Z Test Statistic

In exercises 4–12, let the value of `Statistic` be Z.

4 Let `alpha` = 0.05, D = 50, and `Sample from` be `Normal(0,1)`. Run the lab for the sample sizes 5, 15, 20, 25, and 30. Print each graph. What is happening to the power of the test as the sample size increases?

5 Let n_1 = 10, D = 50, and `Sample from` be `Normal(0,1)`. Run the lab for `alpha` = 0.10, 0.05, 0.025, and 0.01. Print each graph. What is happening to the power of the test as the significance level decreases?

6 Let n_1 = 10, `alpha` = 0.50, and `Sample from` be `Normal(0,1)`. Run the lab for D = 25, 50, 75, and 100. Print each graph. What is happening to the power of the test as the value of D increases?

7 Repeat exercises 4–6 for a `Uniform(0,1)` parent distribution. How do the results of these exercises compare with the results from a Normal(0,1) parent under similar circumstances?

8 Repeat exercises 4–6 for an Exponential(1) parent distribution. How do the results of these exercises compare with the results from a Normal(0,1) parent and from a Uniform(0,1) parent under similar circumstances?

9 Let Sample from be Normal(0,1) and alpha = 0.05. Run the lab for the following combinations of n_1 and D, and print each graph. Which of n_1 or D seems to have the largest effect on the power?

(a) n_1 = 5 and D = 25   (b) n_1 = 25 and D = 25   (c) n_1 = 40 and D = 25
(d) n_1 = 5 and D = 50   (e) n_1 = 25 and D = 50   (f) n_1 = 40 and D = 50
(g) n_1 = 5 and D = 75   (h) n_1 = 25 and D = 75   (i) n_1 = 40 and D = 75

10 Let Sample from be Normal(0,1) and D = 50. Run the lab for the following combinations of n_1 and alpha, and print each graph. Which of n_1 or alpha seems to have the largest effect on the power?

(a) n_1 = 5 and alpha = 0.10   (b) n_1 = 25 and alpha = 0.10
(c) n_1 = 40 and alpha = 0.10   (d) n_1 = 5 and alpha = 0.05
(e) n_1 = 25 and alpha = 0.05   (f) n_1 = 40 and alpha = 0.05
(g) n_1 = 5 and alpha = 0.025   (h) n_1 = 25 and alpha = 0.025
(i) n_1 = 40 and alpha = 0.025   (j) n_1 = 5 and alpha = 0.01
(k) n_1 = 25 and alpha = 0.01   (l) n_1 = 40 and alpha = 0.01

11 Let Sample from be Normal(0,1) and n_1 = 25. Run the lab for the following combinations of alpha and D, and print each graph. Which of alpha or D seems to have the largest effect on the power?

(a) alpha = 0.10 and D = 25   (b) alpha = 0.05 and D = 25
(c) alpha = 0.025 and D = 25   (d) alpha = 0.01 and D = 25
(e) alpha = 0.10 and D = 50   (f) alpha = 0.025 and D = 50
(g) alpha = 0.05 and D = 50   (h) alpha = 0.01 and D = 50
(i) alpha = 0.10 and D = 75   (j) alpha = 0.05 and D = 75
(k) alpha = 0.025 and D = 75   (l) alpha = 0.01 and D = 75

12 For the following combinations of alpha and D, run the lab to find a value of n that will result in a power of at least 0.70, at least 0.80, at least 0.90, at least 0.95, and at least 0.99. How do differing values of alpha, D, and power affect the needed sample size?

(a) alpha = 0.01 and D = 25   (b) alpha = 0.01 and D = 50
(c) alpha = 0.01 and D = 75   (d) alpha = 0.01 and D = 100
(e) alpha = 0.025 and D = 25   (f) alpha = 0.025 and D = 50
(g) alpha = 0.025 and D = 75   (h) alpha = 0.025 and D = 100
(i) alpha = 0.05 and D = 25   (j) alpha = 0.05 and D = 50
(k) alpha = 0.05 and D = 75   (l) alpha = 0.05 and D = 100
(m) alpha = 0.10 and D = 25   (n) alpha = 0.10 and D = 50
(o) alpha = 0.10 and D = 75   (p) alpha = 0.10 and D = 100

16.6.2 The One-sample t Test Statistic

In exercises 13–18, let the value of `Statistic` be `One-sample-t`.

13 Let `alpha` $= 0.05$, `D` $= 50$, and `Sample from` be `Normal(0,1)`. Run the lab for the sample sizes 5, 15, 20, 25, and 30. Print each graph. What is happening to the power of the test as the sample size increases?

14 Let `n_1` $= 10$, `D` $= 50$, and `Sample from` be `Normal(0,1)`. Run the lab for `alpha` $= 0.10, 0.05, 0.025,$ and 0.01. Print each graph. What is happening to the power of the test as the significance level decreases?

15 Let `n_1` $= 10$, `alpha` $= 0.50$, and `Sample from` be `Normal(0,1)`. Run the lab for `D` $= 25, 50, 75,$ and 100. Print each graph. What is happening to the power of the test as the value of D increases?

16 Repeat exercises 13–15 for a `Uniform(0,1)` parent distribution. How do the results of these exercises compare with the results from a Normal(0,1) parent under similar circumstances?

17 Repeat exercises 13–15 for an `Exponential(1)` parent distribution. How do the results of these exercises compare with the results from a Normal(0,1) parent and from a Uniform(0,1) parent under similar circumstances?

18 For the following combinations of `alpha` and D, use the lab to find a value of n that will result in a power of at least 0.70, at least 0.80, at least 0.90, at least 0.95, and at least 0.99. How do differing values of `alpha`, D, and power affect the needed sample size?

(a) `alpha` $= 0.01$ and D $= 25$
(b) `alpha` $= 0.01$ and D $= 50$
(c) `alpha` $= 0.01$ and D $= 75$
(d) `alpha` $= 0.01$ and D $= 100$
(e) `alpha` $= 0.025$ and D $= 25$
(f) `alpha` $= 0.025$ and D $= 50$
(g) `alpha` $= 0.025$ and D $= 75$
(h) `alpha` $= 0.025$ and D $= 100$
(i) `alpha` $= 0.05$ and D $= 25$
(j) `alpha` $= 0.05$ and D $= 50$
(k) `alpha` $= 0.05$ and D $= 75$
(l) `alpha` $= 0.05$ and D $= 100$
(m) `alpha` $= 0.10$ and D $= 25$
(n) `alpha` $= 0.10$ and D $= 50$
(o) `alpha` $= 0.10$ and D $= 75$
(p) `alpha` $= 0.10$ and D $= 100$

16.6.3 The Two-sample t Test Statistic

In exercises 19–24, let the value of `Statistic` be `Two-sample-t`.

19 Let `alpha` $= 0.05$, D $= 50$, and `Sample from` be `Normal(0,1)`. Run the lab for the sample sizes `n_1` $=$ `n_2` $= 5, 15, 20, 25,$ and 30. Print each graph. What is happening to the power of the test as the sample size increases?

20 Let `n_1` $=$ `n_2` $= 10$, D $= 50$, and `Sample from` be `Normal(0,1)`. Run the lab for `alpha` $= 0.10, 0.05, 0.025,$ and 0.01. Print each graph. What is happening to the power of the test as the significance level decreases?

21 Let n_1 = n_2 = 10, alpha = 0.50, and Sample from be Normal(0,1). Run the lab for D = 25, 50, 75, and 100. Print each graph. What is happening to the power of the test as the value of D increases?

22 Repeat exercises 19–21 for a Uniform(0,1) parent distribution. How do the results of these exercises compare with the results from a Normal(0,1) parent under similar circumstances?

23 Repeat exercises 19–21 for an Exponential(1) parent distribution. How do the results of these exercises compare with the results from a Normal(0,1) parent and from a Uniform(0,1) parent under similar circumstances?

24 For the following combinations of alpha and D, run the lab to find a value of n_1 and n_2 that will result in a power of at least 0.70, at least 0.80, at least 0.90, at least 0.95, and at least 0.99. How do differing values of alpha, D, and power affect the needed sample sizes? To complete this assignment, let n_1 = n_2.

(a) alpha = 0.01 and D = 25
(b) alpha = 0.01 and D = 50
(c) alpha = 0.01 and D = 75
(d) alpha = 0.01 and D = 100
(e) alpha = 0.025 and D = 25
(f) alpha = 0.025 and D = 50
(g) alpha = 0.025 and D = 75
(h) alpha = 0.025 and D = 100
(i) alpha = 0.05 and D = 25
(j) alpha = 0.05 and D = 50
(k) alpha = 0.05 and D = 75
(l) alpha = 0.05 and D = 100
(m) alpha = 0.10 and D = 25
(n) alpha = 0.10 and D = 50
(o) alpha = 0.10 and D = 75
(p) alpha = 0.10 and D = 100

16.6.4 The χ^2 Test Statistic

In exercises 25–30, let the value of Statistic be Chi-square.

25 Let alpha = 0.05, D = 50, and Sample from be Normal(0,1). Run the lab for the sample sizes 5, 15, 20, 25, and 30. Print each graph. What is happening to the power of the test as the sample size increases?

26 Let n_1 = 10, D = 50, and Sample from be Normal(0,1). Run the lab for alpha = 0.10, 0.05, 0.025, and 0.01. Print each graph. What is happening to the power of the test as the significance level decreases?

27 Let n_1 = 10, alpha = 0.50, and Sample from be Normal(0,1). Run the lab for D = 25, 50, 75, and 100. Print each graph. What is happening to the power of the test as the value of D increases?

28 Repeat exercises 4–6 for a Uniform(0,1) parent distribution. How do the results of these exercises compare with the results from a Normal(0,1) parent under similar circumstances?

29 Repeat exercises 4–6 for an Exponential(1) parent distribution. How do the results of these exercises compare with the results from a Normal(0,1) parent and from a Uniform(0,1) parent under similar circumstances?

30 For the following combinations of alpha and D, run the lab to find a value of n that will result in a power of at least 0.70, at least 0.80, at least 0.90, at least

0.95, and at least 0.99. How do differing values of alpha, D, and power effect the needed sample size?

(a) alpha = 0.01 and D = 25
(b) alpha = 0.01 and D = 50
(c) alpha = 0.01 and D = 75
(d) alpha = 0.01 and D = 100
(e) alpha = 0.025 and D = 25
(f) alpha = 0.025 and D = 50
(g) alpha = 0.025 and D = 75
(h) alpha = 0.025 and D = 100
(i) alpha = 0.05 and D = 25
(j) alpha = 0.05 and D = 50
(k) alpha = 0.05 and D = 75
(l) alpha = 0.05 and D = 100
(m) alpha = 0.10 and D = 25
(n) alpha = 0.10 and D = 50
(o) alpha = 0.10 and D = 75
(p) alpha = 0.10 and D = 100

16.6.5 The F Test Statistic

In exercises 31–36, let the value of Statistic be F.

31 Let alpha = 0.05, D = 50, and Sample from be Normal(0,1). Run the lab for the sample sizes $n_1 = n_2$ = 5, 15, 20, 25, and 30. Print each graph. What is happening to the power of the test as the sample size increases?

32 Let $n_1 = n_2$ = 10, D = 50, and Sample from be Normal(0,1). Run the lab for alpha = 0.10, 0.05, 0.025, and 0.01. Print each graph. What is happening to the power of the test as the significance level decreases?

33 Let $n_1 = n_2$ = 10, alpha = 0.50, and Sample from be Normal(0,1). Run the lab for D = 25, 50, 75, and 100. Print each graph. What is happening to the power of the test as the value of D increases?

34 Repeat exercises 31–33 for a Uniform(0,1) parent distribution. How do the results of these exercises compare with the results from a Normal(0,1) parent under similar circumstances?

35 Repeat exercises 31–33 for an Exponential(1) parent distribution. How do the results of these exercises compare with the results from a Normal(0,1) parent and from a Uniform(0,1) parent under similar circumstances?

36 For the following combinations of alpha and D, run the lab to find a value of n_1 and n_2 that will result in a power of at least 0.70, at least 0.80, at least 0.90, at least 0.95, and at least 0.99. How do differing values of alpha, D, and power affect the needed sample sizes? To complete this assignment, let $n_1 = n_2$.

(a) alpha = 0.01 and D = 25
(b) alpha = 0.01 and D = 50
(c) alpha = 0.01 and D = 75
(d) alpha = 0.01 and D = 100
(e) alpha = 0.025 and D = 25
(f) alpha = 0.025 and D = 50
(g) alpha = 0.025 and D = 75
(h) alpha = 0.025 and D = 100
(i) alpha = 0.05 and D = 25
(j) alpha = 0.05 and D = 50
(k) alpha = 0.05 and D = 75
(l) alpha = 0.05 and D = 100
(m) alpha = 0.10 and D = 25
(n) alpha = 0.10 and D = 50
(o) alpha = 0.10 and D = 75
(p) alpha = 0.10 and D = 100

17

Calculating One-way ANOVA

Because StataQuest does not provide a method for calculating results for a one-way ANOVA when you have only the summary statistics $\bar{X}_1, \ldots, \bar{X}_K$ and $s_1^2, \ldots, s_K^2$, we provide a lab that will do this for $2 \leq K \leq 6$.

17.1 Introduction

Traditional one-way ANOVA assumes that we have K random samples from normally distributed populations all having the same variance σ^2 but possibly differing means. There are two ways this typically arises. First, we might simply have a set of samples from several populations. Often, however, we have the samples as a result of a completely randomized experiment in which we have $N = n_1 + \cdots + n_K$ "experimental units" and K "treatments" and we randomly assign treatment 1 to n_1 of the units, treatment 2 to n_2 of the units, and so on.

In either case, to test the null hypothesis that all of the population means are the same, we calculate the F statistic

$$F = \frac{MSB}{MSW},$$

where $MSB = SSB/df_B$, $MSW = SSW/df_W$, $df_B = K - 1$, $df_W = n - K$, and

$$SSB = \sum_{i=1}^{K} n_i(\bar{X}_i - \bar{X})^2, \qquad SSW = \sum_{i=1}^{K} (n_i - 1)s_i^2.$$

MSB and MSW are the *mean squares* between and within, SSB and SSW are the *sum of squares* between and within, and df_B and df_W are the *degrees of freedom* between and within. Note that MSW is the K sample generalization of the two-sample pooled estimate of the common variance of the normal populations.

We can then find the p value of the test by finding the area under the $F_{K-1,n-K}$ curve to the right of the F statistic.

17.2 Objectives

This lab is primarily computational. It provides a means of doing one-way ANOVA when we have only the sample sizes, means, and variances (or standard deviations) for a set of random samples.

17.3 Description

To start the lab, select `Calculating One-Way ANOVA` from the `Labs` menu. The dialog box shown in Figure 17.1 will appear giving information about the lab. The box also contains these items:

1. A box labeled `#Treats`, where you specify the number of samples being analyzed. The initial value is 2, and the only choices are 2, 3, 4, 5, and 6.
2. A box where you specify whether variances `s^2` or standard deviations `s` are to be entered.
3. A series of boxes where you enter the sample sizes, means, and either variance or standard deviations.

FIGURE 17.1
Dialog box for the
One-way ANOVA Lab.

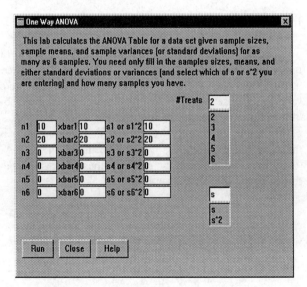

4 The usual three buttons:
   (a) `Run`: After you fill in the required information in the dialog box, clicking on Run executes the lab. A results window displays the sample sizes, means, standard deviations, and variances of the individual samples. Then a line is displayed containing the sample mean of all of the elements of the samples, the ratio of the largest standard deviation to the smallest, and the pooled estimate of the sample variance. Finally, a standard ANOVA table is displayed.
   (b) `Close`: Clicking on this button closes the lab and returns you to the Stat-Concepts menus.
   (c) `Help`: Clicking on this button opens a help window containing information about the lab.

17.4 Guided Tour of the Lab

Because this lab is primarily computational and not conceptual, our guided tour will consist of using the lab to analyze a data set.

17.4.1 The Data

The sample means and variances of the *critical flicker frequency* for people's eyes for random samples of people having three different iris colors are given by

```
Color      n_i      Mean     Variance
-------------------------------------
Brown       8       25.59     1.86
Green       5       26.92     3.40
Blue        6       28.17     2.33
-------------------------------------
```

Does the average flicker time depend on iris color?

17.4.2 Analysis of the Example

We must click on 3 treatments and enter the sample sizes, sample means, and sample variances, as well as on `s^2` because we have entered variances and not standard deviations.

The output given in the results window is shown in Table 17.1. It begins with a summary of the input data and then the pooled mean and pooled standard deviation and the ratio of the largest standard deviation to the smallest. A popular rule of thumb is that the equal variance assumption in the ANOVA is reasonable if that ratio is less than 2. In our example, the value is 1.352, so we continue to the ANOVA table, which appears below the output. Because the p value for the test of equality of means is 0.0231, which is less than 0.05, we conclude that the means are, in fact, significantly different and that average flicker time does, in fact, depend on iris color.

TABLE 17.1
Output of the *Calculating One-way ANOVA* Lab for the example.

```
                    Input Data
Sample     n_i       Mean         Std Dev        Variance
----------------------------------------------------------
  1         8      25.590000      1.363818       1.860000
  2         5      26.920000      1.843909       3.400000
  3         6      28.170000      1.526434       2.330000
----------------------------------------------------------
Overall Mean = 26.754737, s_max/s_min = 1.3520196, s_P = 1.5465688


            Analysis of Variance
Source     df         SS            MS            F         Prob > F
--------------------------------------------------------------------
Between     2      23.007274     11.503637     4.809464      0.0231
Within     16      38.270000      2.391875
--------------------------------------------------------------------
Total      18      61.277274      3.404293
```

17.5 Summary

This lab makes it easy to perform one-way ANOVA when given only the summary statistics of a data set.

18

Between and Within Variation

The F test for comparing the means of several normal populations is based on the idea of comparing how much variability there is between the sample means of the samples to how much variability there is within the individual samples.

18.1 Introduction

Suppose we have a sample of size n_1 from one population, one of size n_2 from another, and so on, for a total of K samples. We assume that each sample is from a normally distributed population, all having the same variance σ^2 but with possibly differing means.

Let $\bar{X}_1, \ldots, \bar{X}_K$ denote the sample means of the K samples and $s_1^2, \ldots, s_K^2$ denote the sample variances. We are seeking a method for determining whether it is reasonable to conclude that the means of the populations are the same.

If the populations all have the same mean, we can expect the sample means to be approximately the same. Thus, if they are "quite different," we should conclude that the population means are different. The difficulty is in determining how much variation in the sample means is too much.

We need a measure of how much variability there is in the $\bar{X}$'s. One measure that statisticians have shown to be effective is the *sum of squares between samples*,

$$SSB = \sum_{i=1}^{K} n_i (\bar{X}_i - \bar{X})^2,$$

where $\bar{X}$ is the *overall mean*, that is, the mean of all of the data. This formula calculates how far each sample mean is from the mean of all the data by squaring this distance and then multiplying by the sample size for that sample. We do this multiplication so that if a large sample has a mean far away from the overall mean, we can give it more weight. Then we add together all of these weighted distances.

If the K sample means are close together, then each one will be close to the overall mean and SSB will be small. But if there is a lot of variability in the sample means, some of them will be far away from the overall mean and SSB will be large. Thus, we want to conclude that the population means are probably different if SSB is "large" and probably not different if SSB is "small."

As with the correlation coefficient and other statistics, we want a measure of variation of the means that is independent of the units used to measure the data. Thus, rather than use SSB directly, we transform it by dividing by an estimate of σ^2. Therefore, we use the F statistic

$$F = \frac{MSB}{MSW},$$

where $MSB = SSB/(K-1)$ and $MSW = SSW/(N-K)$ are the *mean squares for between and within samples* and SSW is the *sum of squares within samples*:

$$SSW = \sum_{i=1}^{K} (n_i - 1)s_i^2.$$

Note that N is the total sample size, that is, N is the sum of the n_i's.

We can find the p value of the test for equality of population means by finding the area under the $F_{K-1, N-K}$ curve to the right of the F statistic.

18.2 Objectives

1. Understand that as the variability within samples increases, the F test has more difficulty determining that means are different.
2. Understand that the power of the F test increases as the sample sizes increase.

18.3 Description

To start the lab, select Between and Within Variation from the Labs menu. The dialog box shown in Figure 18.1 will appear containing information about the lab, as well as a box labeled n and five buttons that we will describe shortly.

A graphics window also appears similar to the one shown in Figure 18.2. This window is the result of the lab generating two sets of four random samples. Each of the eight samples has the same number $n = 10$ of elements (this is specified in the box labeled n). Each sample is from a normal population having the same variance $\sigma^2 = 1$. The first four populations have different means, namely, 17, 22, 17, and 22. The second four populations all have the same mean, namely, 20.

FIGURE 18.1
Dialog box for the *Between and Within Variation* Lab.

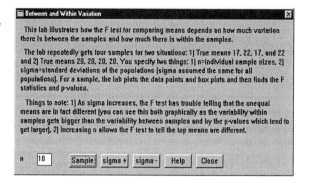

FIGURE 18.2
Result of the lab for $\sigma = 1$ and $n = 10$.

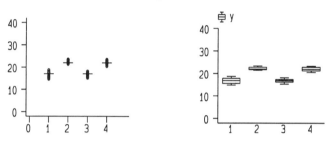

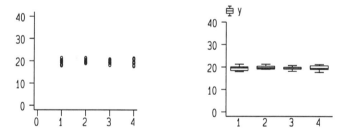

The top half of the graphics window plots the first set of four samples (the ones with population means 17, 22, 17, and 22) both with side-by-side dotplots and side-by-side box plots. Red horizontal lines are placed on the dotplots to show where the population means are located. With $\sigma^2 = 1$, there is very little variability within samples.

The bottom half of the window shows similar plots but for the second set of four samples, the ones from populations all having mean 20.

The values of n and σ and the F statistic and its p value are displayed above each set of plots.

These five buttons and a box appear at the bottom of the dialog box:

1 Sample: Clicking on this button generates a new set of eight samples and produces the same sort of display. The new samples are produced under the same conditions as the previous one.

2. `sigma +`: Clicking on this button increases the previous value of σ by 1 and generates a new set of samples. The initial value of σ is 1, and σ must be a number between 1 and 6. Clicking on `sigma +` after $\sigma = 6$ results in the lab using the value 6.

3. `sigma -`: Clicking on this button decreases the previous value of σ by 1 and generates a new set of samples. Clicking on `sigma -` after $\sigma = 1$ results in the lab using the value 1.

4. `Help`: Clicking on this button opens a help window containing information about the lab.

5. `Close`: Clicking on this button closes the lab and returns you to the StatConcepts menus.

6. `n`: This box is where you can specify how large you want each of the eight samples to be. The initial value is 10, and `n` must be a number between 2 and 100.

18.4 Guided Tour of the Lab

In our tour, we will refer to the top set of samples as the *unequal means case* and the bottom set as the *equal means case* because in the top set, the true means are unequal, and in the bottom set, they are equal.

18.4.1 Level of Significance and Power of the F Test

The probability of rejecting equality of means in the top set is called the *power* of the test because it is how likely the test is to reject the null hypothesis when, in fact, it should be rejected. In the bottom half, the probability of rejection is the *level of significance* or probability of a Type I error because we don't want to reject it in this case.

Recall that in any hypothesis test, we reject the null hypothesis at level α if the p value of the test is less than α.

We start our tour by looking further at Figure 18.2. The two F statistics are 95.50 and 0.56, with corresponding p values of 0.000 (so we correctly reject in the unequal means case) and 0.643 (so we correctly don't reject in the equal means case). This is good! Start the lab and see if you get similar results. Now click on `Sample 20 times` and keep a record of how many times you reject equality of means in both the equal means case and the unequal means case. When we did it, we correctly rejected equality every time in the unequal means case and correctly failed to reject 18 of the 20 times in the equal means case. What did you get?

18.4.2 The Effect of Variation Within Samples

Now let's run the lab by clicking on `sigma +` five times so that σ becomes 6. Figure 18.3 shows the graph for our first set of samples when $\sigma = 6$. Notice that for this set of eight samples, we don't reject equality of means for either case (the p values are 0.200 and 0.673). Thus, in the unequal means case, we have made a Type II error, that is, we failed to reject the null hypothesis (equal means) when it is false. Now click on `Sample 20 times`, again keeping track of the number of

FIGURE 18.3
Result of the lab for $\sigma = 6$ and $n = 10$.

True Means Are 17 22 17 22, n=10, sigma=6, F= 2.96, p-value=0.200

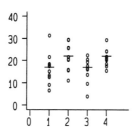
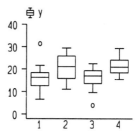

True Means Are 20 20 20 20, n=10, sigma=6, F= 0.57, p-value=0.673

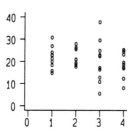
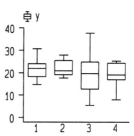

rejections. When we did it, we only got rejections in the unequal means case four times (which gives us power of around $4/20 = 0.2$ or 20%) and rejections in the equal means case only once (which is perfect because we incorrectly rejected the true hypothesis $1/20 = 0.05$ or 5% of the time).

Therefore in both the $\sigma = 1$ and $\sigma = 6$ cases, our Type I error probability in the equal means case was around 0.05, as hoped, but the power in the unequal means case for the two values of σ dropped from 100% to 20%. What caused this to happen? The answer is found in comparing Figures 18.2 and 18.3. In the first figure, looking at either the top dotplot or box plot, *because of the lack of much variability* within each sample, it is obvious that the means are different. On the other hand, it is difficult to see this in Figure 18.3 because the individual samples are so spread out that they overlap quite a bit. This causes the F test to be unable to tell that the means are different.

We have only looked at $\sigma = 1$ and $\sigma = 6$. In the chapter exercises, you will look at the values in between.

18.4.3 The Effect of Sample Size

Now we'll do a similar experiment for $\sigma = 6$ and larger n to see what happens to the power of the test in the unequal means case. Recall that for $n = 10$, the power was only around 20%.

Change n to 20 and use sigma + if necessary so that $\sigma = 6$. Figure 18.4 shows one result for these values. Notice that the individual samples are still rather spread out, but the fact that there are more values in each sample makes it easier to see the difference in location of the samples. Now click on Sample 20 times and count

FIGURE 18.4
Result of the lab for $\sigma = 6$ and $n = 20$.

True Means Are 17 22 17 22, n=20, sigma=6, F= 2.90, p-value=0.208

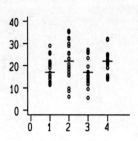 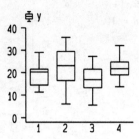

True Means Are 20 20 20 20, n=20, sigma=6, F= 0.63, p-value=0.645

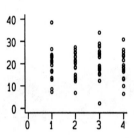 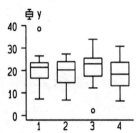

the numer of rejections in the unequal means case. We got seven rejections, so our estimate of power went up from 20% to 35% ($7/20 = 0.35$). Now let $n = 30$ and do it again (Figure 18.5 shows a typical output). Remarkably, when we did it, our power went up to 80% because we got only four rejections.

In the chapter exercises, you will do this experiment for other values of n and σ.

18.5 Summary

In the brief tour, we saw several sights:

1. The level of significance of the F test is the probability of rejecting the hypothesis of cqual population means when they are all actually equal. This probability is unaffected by sample size or variability within samples.

2. The power of the F test is the probability of rejecting the hypothesis of equal population means when they are actually different. This probability is greatly affected by both sample size and degree of variability within samples.

18.6 Lab Exercises

1. Repeat the process used in Subsection 18.4.2 for each value of σ from 1 to 6, and make a table of the proportion of times equality of means is rejected in the unequal and equal means cases. Does σ matter in the equal means case? Use the

FIGURE 18.5 Result of the lab for $\sigma = 6$ and $n = 30$.

True Means Are 17 22 17 22, n=30, sigma=6, F= 7.80, p-value=0.010

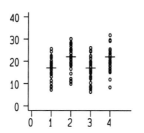
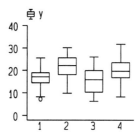

True Means Are 20 20 20 20, n=30, sigma=6, F= 0.20, p-value=0.893

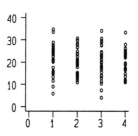
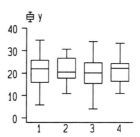

StataQuest editor to enter the proportions for the unequal means case (as well as the corresponding values of σ) into a StataQuest data set, and then do a plot of the proportion versus σ. Use Graphs→ Time series→ Plot Y vs. X to do the graph.

2 For $\sigma = 5$, repeat the process used in Subsection 18.4.3 for $n = 5, 10, 15, 20, 25,$ and 30, and make a table of the proportion of times equality of means is rejected in the unequal and equal means cases. Does n seem to matter in the equal means case? Using the StataQuest editor, enter the proportions for the unequal means case (as well as the corresponding values of n) into a StataQuest data set and then do a plot of the proportion versus n. Use Graphs→ Time series→ Plot Y vs. X to do the graph.

19

Chi-square Goodness of Fit

The chi-square goodness-of-fit test is an important method for testing whether a sample comes from a certain distribution. In this lab, we will illustrate this by testing whether a set of points is uniformly distributed in a square.

19.1 Introduction

Points are said to be uniformly distributed in a geometric region (square, circle, and so on) if the probability that a point falls in some part of the region is equal to the area of that part of the region divided by the area of the entire region. In this lab, we will be generating points uniformly distributed in a square whose sides are of length 1. Notice that this is the analog on the $X - Y$ *plane* of the uniform distribution on an *interval*, which we have used extensively throughout this book. Recall the histogram of the Uniform(0,1) sample in the *How Are Populations Distributed?* Lab. The analog in this chapter would be to count how many points fall in little squares in the plane and put a solid bar above each little square whose height is proportional to how many points are in that square.

Given a set of n points, we divide the square into subsquares (or bins) and count how many of the points are in each bin. We let O_{ij} denote the number of points in the ith row and jth column of bins. If the n points are uniformly distributed in the square, then we would expect the number of points in each bin to be n/M, where M

is the total number of bins. This expected number of points in each square is denoted by $E_{ij} = n/M$. Thus,

$$\chi^2 = \sum_{ij} \frac{(O_{ij} - E_{ij})^2}{E_{ij}}$$

is a measure of how close the observed counts are to the expected counts. Assuming that the null hypothesis of uniformity is true, if the expected count in each bin is at least 5, then the χ^2 statistic has approximately the χ^2 distribution with $M - 1$ degrees of freedom, and we can find the p value of the test by finding the area under the χ^2_{M-1} curve to the right of the value of χ^2 for the observed set of points.

Recall that for any test of a hypothesis, we reject the null hypothesis at level of significance α if the p value of the test is less than α.

19.2 Objectives

1. Show what uniformity looks like in a geometric region.
2. Show how the χ^2 goodness-of-fit test works in situations like this one.

19.3 Description

Clicking on Chi-square Goodness of Fit from the Labs menu brings up the dialog box shown in Figure 19.1, which briefly describes the lab. The box also contains two items on the left labeled n and bins and three buttons:

1. Run: Clicking on this button results in the lab producing a display in the graphics window similar to the one shown in Figure 19.2. The lab uniformly selects a set

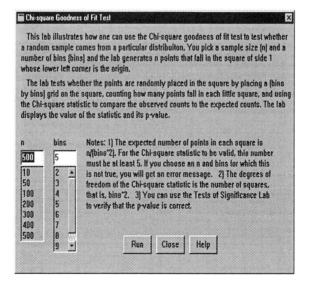

FIGURE 19.1
Dialog box for the *Chi-square Goodness of Fit* Lab.

FIGURE **19.2**
Result of the lab using
$n = 500$ and $\text{bins} = 5$.

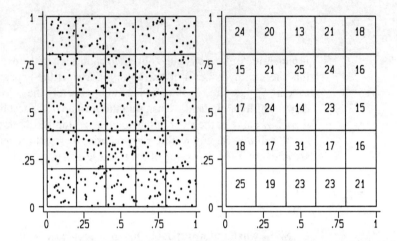

of points from a square and plots them. The number of points is the value you select in the n item. Then the lab draws a set of boxes, or bins, on top of the square and counts the number of points in each box. The number of boxes in each row of boxes is specified in the bins item. The lab then draws another set of boxes to the right of the original set of boxes and writes in each box how many points are in that box. Finally, the lab calculates and writes at the top of the graphics window the value of the χ^2 statistic and the p value of the test for uniformity.

2 Close: Clicking on this button closes the lab and returns you to the StatConcepts menus.

3 Help: Clicking on this button opens a help window containing information about the lab.

4 n: Here you can choose the number of points to generate within the square. The initial value is 500, and the only restriction is that n divided by the square of bins must be at least 5.

5 bins: Here you specify the number of bins in each row in the square. The initial value is 5, and the only restriction on the value is that n divided by the square of bins must be at least 5.

19.4 Guided Tour of the Lab

We start our tour by looking more closely at Figure 19.2. The lab has generated 500 points uniformly in the square, placed a 5-by-5 "grid" on top of the square, and counted how many points are in each of the grid boxes. Thus, we used 5 for bins and 500 for n. Because there are 500 points and 25 bins, we would expect $500/25 = 20$

points in each bin. Looking through the list of counts, we see that the smallest is 13 and the largest is 25. This seems to be a rather large deviation from having 20 in each box but, in fact, it is not! The value of χ^2 is 22.60, which for 24 degrees of freedom gives a p value of 0.54. Thus, we do not reject the null hypothesis of uniformity.

Look again at the scatter of points. Note that this is similar to looking at where stars are in the sky. In fact, one use of the method of this chapter is to answer the question "Are stars uniformly distributed in the sky?" In the figure, there appear to be too many empty places, but again, we know that these points are uniformly distributed. You should run the lab several times with these values of n and bins to see what a uniform distribution looks like in a square.

Next, run the lab for n = 200 and bins = 2 so that the big square has been divided into four smaller squares. The result is shown in Figure 19.3. The expected number of points in each bin is 50. We selected this possible result because it was one for which the null hypothesis is rejected (the p value is 0.02). Thus, this is one for which the test says the points are nonuniform.

The last part of the tour involves trying out a wide variety of combinations of n and bins to learn what uniformity means.

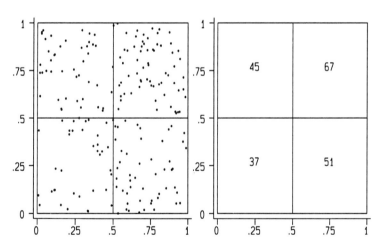

FIGURE 19.3
Result of the lab using n = 200 and bins = 2.

19.5 Summary

On this last leg of our tour, we learned several things:

1. A set of points can look surprisingly nonuniform even when they are, in fact, being produced from a uniform population.
2. The χ^2 test is used to test the "goodness" of a conjectured distribution (such as uniformity in a square).

19.6 Lab Exercises

1. Which values of n can be used for bins = 5? How about bins = 10?
2. Run the lab for n = 500 and bins = 2, 5, and 10. What are the chi-square and p values? Why did the chi-square values increase?
3. For n = 300 and bins = 5, find a set of points that the lab judges to be nonuniform (for $\alpha = 0.05$) and print the graph. What features of the scatter of points and the counts in the bins led the procedure to conclude the points were nonuniform?
4. Run the lab 20 times for each of the following combinations of n and bins: (50,3), (100,4), (200,5), (300,6), (400,8), and (500,10). Each time, keep track of the number of times the null hypothesis of uniformity was rejected. How many times would you expect it to be rejected? Does sample size seem to matter in how often rejection occurs?
5. How would we have to change the lab so that we could study the power of the test in terms of n and bins?

INDEX

0–1 population 18, 57, 106, 136

A
approximating binomial probabilities 58, 72 (*cf*), 136–138
alternative hypothesis 207–209, 215, 255
ANOVA 289 (*cf*), 293 (*cf*)
ASCII data set 4

B
bell-shaped curve 33, 101, 102, 106, 114, 129
best estimator 170
beta distribution 41–42, 54
between variation 293 (*cf*)
binomial distribution 58, 59, 65 (*cf*), 74, 81
bivariate data set 83
bivariate descriptive statistics 83 (*cf*)
bivariate sampling schemes 83–84
boxplot 172

C
calculator 73–74
center of a distribution 170
Central Limit Theorem 36, 101 (*cf*), 137, 150–151, 171
Cauchy distribution 42–43, 54
χ^2 distribution 38–39, 54, 114, 130–131, 159
χ^2 goodness of fit 300 (*cf*)
χ^2 statistic 38, 113 (*cf*), 145, 146, 159–162, 211, 207, 231–233, 252
χ^2 test 231–233, 275–279

confidence intervals 182 (*cf*), 198 (*cf*), 200–205
  difference in two population means $\mu_1 - \mu_2$ 184, 193–194
  difference in two population proportions $\pi_1 - \pi_2$ 185, 195
  population mean μ 184, 188–191
  population proportion π 184, 194
  population variance σ^2 184
  ratio of two population variances σ_1^2/σ_2^2 184, 194
continuous distributions 54–56
  beta 41–42, 54
  Cauchy 42–43, 54
  χ^2 38–39, 54, 114, 130–131, 159
  exponential 44–45, 54, 105–106
  F 38, 40–41, 54, 114, 132–134
  gamma 45–46, 55
  Laplace 46, 55
  logistic 46–47, 55
  lognormal 47–48, 55
  normal 33, 35, 36–37, 55, 74, 102, 105, 114, 129, 130
  Pareto 48–49, 55
  Student t 37, 55, 114, 126–127, 134–136
  uniform 50–51, 56, 106
  Weibull 51–52, 56
continuous population 18
consistent 102
correlation coefficient 84–86, 92–94
critical values 115, 125–129

D
data set in *StatConcepts* 4

degrees of freedom
  between (in ANOVA) 289–290
  χ^2 distribution 39, 54, 114
  χ^2 statistic 146
  denominator 40, 54, 114, 133–134, 147
  effect on critical values of Student t distribution 126–127
  F distribution 40, 54, 114, 132–134
  F statistic 147
  numerator 40, 54, 114, 132–133, 147
  one-sample t statistic 146
  paired t statistic 205, 254
  Student t distribution 37, 55, 114
  two-sample pooled t statistic 146
  two-sample t statistic, unequal population variances 205
  within (in ANOVA) 289–290
description of labs
  Approximating Binomial Probabilities 62–64
  Between and Within Variation 294–296
  Bivariate Descriptive Statistics 88–92
  Calculating Confidence Intervals 198–200
  Calculating One-Way ANOVA 290–291
  Calculating Tests of Hypotheses 244–246
  Central Limit Theorem 102–104
  Chi-square Curves 119–120

Chi-square Goodness of Fit 301–302
Critical Values 116–117
F Curves 121–122
How Are Populations Distributed? 34–35
Interpreting Confidence Intervals 185–187
Introduction to Concept Labs 20
Least Squares 90–92
Level of Significance of a Test 217–219
Minimum Variance Estimation 172–174
Normal Approximation to Binomial 63–64, 123–124
Normal Curves 118–119
Poisson Approximation to Binomial 62–63
Power of a Test 258–260
The Negative Binomial Distribution Lab 61–62
Random Sampling 22–24
Relative Frequency and Probability 28–30
Sampling Distributions 147–149
Sampling From 0–1 Populations 60–64
Sampling With and Without Replacement 60–64
Scatterplots I 88–89
Scatterplots II 89–90
t Converging to Z 122–123
Tests of Significance 209–211
Z, t, χ^2, F 116–124
discrete distributions 24, 58 (*cf*), 81–82
  binomial 58, 59, 65 (*cf*), 74, 81
  hypergeometric 58, 59, 65 (*cf*), 82
  Poisson 72, 82
  negative binomial 59, 70 (*cf*), 82
  uniform 24
discrete population 24
discrete uniform distribution 24
distribution curve 33
distribution family 34
distribution of population 22, 33 (*cf*)

E
editor 4, 5–6
Empirical Rule 102, 106, 107
exiting *StatConcepts* 2

expected value 58
exponential distribution 44–45, 54, 105–106
exponential(1) parent 105–106
  effect on confidence interval for population mean μ 190
  effect on confidence interval for population variance σ^2 192
  effect on level of significance of χ^2 test 233
  effect on level of significance of F test 235–236
  effect on level of significance of one-sample t test 226–227
  effect on level of significance of two-sample pooled t test 230
  effect on level of significance of Z test 223–224
  effect on power of χ^2 test 278–279
  effect on power of F test 282
  effect on power of one-sample t test 271
  effect on power of Z test 264–265
  effect on sampling distribution of χ^2 statistic 161–162
  effect on sampling distribution of F statistic 163–165
  effect on sampling distribution of one-sample t statistic 155
  effect on sampling distribution of two-sample pooled t statistic 158–159
  effect on sampling distribution of Z statistic 150

F
failure 57
family of a distribution 34
F distribution 38, 40–41, 54, 114, 132–134
F statistic 40, 113 (*cf*), 145, 147, 162–165, 207, 234–236, 252
  in ANOVA 289, 294
F test 234–236, 252, 279–283
finite population 18
finite population correction factor 59, 68, 69
fitted value 87
fixed X 83–84, 92
formulas for calculating confidence intervals 204–205, 251–254

formulas for calculating tests of hypotheses 251–254

G
gamma distribution 45–46, 55
generating random numbers 18 (*cf*)

H
help in *StatConcepts* 5
histogram 3–4, 172
hypergeometric distribution 58, 59, 65 (*cf*), 81
hypothesis test 215–216, 224 (*cf*), 246–250, 255

I
independence 18
independent samples 24
inferential statistics 34
installing *StatConcepts* 1–2
intercept 86

L
Laplace distribution 46, 55
least squares intercept 86
least squares regression 84–87, 94–98
least squares slope 86, 95–96
left-sided alternative 208, 211
level α critical value 115
level of significance 215 (*cf*), 256, 257
  ANOVA 296
  effect of type of test for Z test 220
  effect of number of bins for χ^2 goodness of fit test 302–303
  effect of sample size for ANOVA F test 297–298
  effect of within variation for ANOVA F test 296–297
  effect on power of χ^2 test 277
  effect on power of F test 280
  effect on power of two-sample pooled t test 273
  effect on power of Z test 262
logistic distribution 46–47, 55
lognormal distribution 47–48, 55
lottery 27, 31
lower adjacent value 172

M
meaning of confidence 183, 191, 193
mean squares between 289–290, 294

Index

mean squares within 289–290, 294
minimum variance 170, 171–172
minimum variance estimation 170 (*cf*)
   of center of Laplace(0,1) distribution 177–178
   of center of standard normal distribution 175–176
   of center of Student t_3 distribution 178–179
   of center of uniform(−0.5, 0.5) distribution 177
modeling 34
modulus 18
multiplier 18
multiple correlation coefficient 87, 96–98

N
negative binomial distribution 59, 70 (*cf*), 82
noncentrality parameter 268
non-central t distribution 267
normal(0,1) parent 106
normal approximation to binomial distribution 58, 75–78, 136–138
normal distribution 33, 35, 36–37, 55, 74, 102, 105, 114, 129, 130
null hypothesis 215, 255

O
one-sample t statistic 37, 113 (*cf*), 144, 146, 151–157, 206–207, 225–228, 251
one-sample t test 225–228, 251 266–271
one-way ANOVA 289 (*cf*), 293 (*cf*)
overall mean 294

P
paired data 205, 253
parameter estimation 34
parametric inference 34
parent distribution 36, 105
Pareto distribution 48–49, 55
percentile 148, 150
Poisson approximation to binomial probabilities 58, 72–73
Poisson distribution 72, 74–75
pooled sample variance 207, 229
pooled t statistic 145, 146, 157–159, 207, 228–230, 253
pooled t test 228–230, 272–275, 253

poplation 21, 105
population distributions 22, 33 (*cf*)
power of the test 216, 255 (*cf*)
   ANOVA 296
   effect of distance between true and hypothesized value for Z test 262
   effect of level of significance for Z test 262
   effect of sample size for ANOVA F test 297–298
   effect of sample size for Z test 262
   effect of within variation for ANOVA F test 296–297
   interpretation 256
   normal theory power 257
   sample size determination 257, 265–266
probability 27 (*cf*)
probability of an event 28, 30
probability of a sequential pair 31
p-value 207 (*cf*), 216, 251–254, 256

R
random sample 28
random sampling 21 (*cf*)
random X 83
rejection region 216, 251–254, 256
relative frequency 27 (*cf*)
residuals 87
right-sided alternative 207, 211

S
sample 21
sample mean 170
sample median 170
sample proportion 137
samples from uniform(0,1) populations 18
sampling from a 0–1 population 57 (*cf*)
sampling from a continuous population 19
sampling distribution 144 (*cf*)
   of a statistic 22, 113, 114, 145
   of the χ^2 statistic 146, 159–162, 207
   of the F statistic 147, 162–165, 207
   of the one-sample t statistic 146, 151–157, 207

   of the sample mean 24–25, 36, 101 (*cf*), 133, 145
   of the two-sample pooled t statistic 146, 157–159, 207
   of the Z statistic 145, 149–151, 207
sampling distribution of the sample mean $\bar{X}$ 101 (*cf*)
sampling with replacement 19, 57, 58, 59, 65, 70
sampling without replacement 19, 27, 58, 59, 65
sample 21
scatterplots 84 (*cf*)
seed 18
simple random sample 22
slope 86, 95–96
standard normal distribution 106, 114, 125–127, 129, 135
starting *StatConcepts* 2
StataQuest 2 (*cf*)
StataQuest tool bar 7–8
StataQuest windows 6–7
StatConcepts menus 8–17
statistic 22, 113
   χ^2 145, 146, 207, 211, 231–233
   F 145, 147, 207, 212, 234–236
   formulas for calculating 251–254
   one-sample t 144, 146, 206, 211, 225–228
   two-sample pooled t 144, 207, 211, 228–230
   Z 144, 145, 149, 206, 211, 219–224
Student t distribution 37, 55, 114, 126–127, 134–136
success 57
summaries menu 13–15
sum of exponential random variables 45
sum of squared independent standard normal random variables 146
sum of squares between 289–290, 293–294
sum of squares within 289–290, 294

T
tails of a distribution 43
tails of a Cauchy distribution 42
t distribution 37, 55, 114, 125–127, 134–136
tests of significance 206 (*cf*)

test statistic 215–216
t statistic 37, 113 (cf), 144, 146, 206–207, 151–157, 225–228, 251
t test 225–228, 251, 256, 266–271
two-sample pooled t statistic 145, 146, 157–159, 207, 228–230, 253
two-sample pooled t test 228–230, 272–275, 253
two-sided alternative 208–209, 212
Type I error 216, 256, 257, 296
Type II error 216, 256, 296
types of tests 207, 208–209, 220, 222–224

U
unbiased estimator 101, 170–172
unbiased minimum variance estimator 171
unequal means case in ANOVA 296
uniform(0,1) parent 105
    effect on level of significance of χ^2 test 231
    effect on level of significance of F test 234–235
    effect on level of significance of one-sample t test 225–226
    effect on level of significance of two-sample pooled t test 229–230
    effect on level of significance of Z test 220–223
    effect on power of χ^2 test 278
    effect on power of F test 281–282
    effect on power of one-sample t test 269–271
    effect on power of Z test 263
    effect on sampling distribution of χ^2 statistic 160
    effect on sampling distribution of F statistic 162–163
    effect on sampling distribution of one-sample t statistic 152–155
    effect on sampling distribution of two-sample pooled t statistic 157
    effect on sampling distribution of Z statistic 150
uniform distribution 50–51, 56, 106
uniformity 18
uniformly distributed on an interval 300
uniformly distributed on a geometric region 300
upper adjacent value 172

V
variable window 7

W
Weibull distribution 51–52, 56
whiskers 172
within variation 293 (cf)

Z
Z statistic 113 (cf), 145, 149–151, 206, 219–224, 251, 260–262
Z test 219–224, 251, 260–266